Springer Series in **Materials Science**

Edited by Ulrich Gonser

D0911637

Springer
Berlin
Heidelberg
New York
Barcelona
Budapest
Hong Kong
London
Milan
Paris
Tokyo

Springer Series in *Materials Science*

Advisors: M.S. Dresselhaus · H. Kamimura · K.A. Müller
Editors: U. Gonser · R.M. Osgood · M.B. Panish · H. Sakaki
Managing Editor: H.K.V. Lotsch

Martin von Allmen   Andreas Blatter

Laser-Beam Interactions with Materials

Physical Principles and Applications

Second, Updated Edition

With 78 Figures

Springer

Dr. Martin von Allmen

Dr. GRAF AG
CH-4563 Gerlafingen
Switzerland

Dr. Andreas Blatter

SM-Thun
Allmendstrasse 74
CH-3600 Thun
Switzerland

Series Editors:

Prof. Dr. U. Gonser

Fachbereich 12.1, Gebäude 22/6
Werkstoffwissenschaften
Universität des Saarlandes
D-66041 Saarbrücken, Germany

M. B. Panish, Ph. D.

AT&T Bell Laboratories
600 Mountain Avenue
Murray Hill, NJ 07974-2070, USA

Prof. R. M. Osgood, Jr.

Microelectronics Science Laboratory
Department of Electrical Engineering
Columbia University
Seeley W. Mudd Building
New York, NY 10027, USA

Prof. H. Sakaki

Institute of Industrial Science
University of Tokyo
7-22-1 Roppongi, Minato-ku
Tokyo 106, Japan

Managing Editor:

Dr.-Ing. Helmut K. V. Lotsch

Springer-Verlag, Tiergartenstrasse 17
D-69121 Heidelberg, Germany

ISBN 3-540-59401-9 Springer-Verlag Berlin Heidelberg New York

ISBN 3-540-17568-7 1. Auflage Springer-Verlag Berlin Heidelberg New York
ISBN 0-387-17568-7 1st Edition Springer-Verlag New York Berlin Heidelberg

Library of Congress Cataloging-in-Publication Data.

Von Allmen, M. Laser -beam interactions with materials: physical principles and applications / Martin von Allmen, Andreas Blatter. – 2nd updated ed. p. cm. – (Springer series in materials science; v.2) Includes bibliographical references and index. ISBN 3-540-59401-9 (softcover: alk. paper) 1. Materials-Effect of radiation on. 2. Laser beams-Industrial applications. I. Blatter, Andreas, 1960 - . II. Title. III. Series. TA418.6.V63 1995 620.1'1228– dc20 95-32586

Typesetting: PS™ Technical Word Processor
SPIN: 10489808   54/3144-5 4 3 2 1 0 – Printed on acid-free paper

Preface

Eight years after the appearance of the first edition of *Laser-Beam Interactions with Materials*, the topic seems far from losing any of its vigour and fascination - if the rate of papers published is any indication. A number of interesting new applications have appeared in the meantime, and a tremendous amount of work on process characterisation has been done. Nevertheless, the main ideas of eight years ago are still there, and so are some of the old puzzles.

This second edition, which owes its existence to the friendly reception of the first one, comes with a number of corrections, updates and timely additions (notably a new section on laser deposition) but preserves both the original layout of the monograph and its emphasis on the physics behind the phenomena.

Bern, September 1994

M. von Allmen
A. Blatter

Preface to the First Edition

Lasers, having proven useful in such diverse areas as high-resolution spectroscopy and the guiding of ferryboats, are currently enjoying great popularity among materials scientists and engineers. As versatile sources of "pure" energy in a highly concentrated form, lasers have become attractive tools and research instruments in metallurgy, semiconductor technology and engineering. This text treats, from a physicist's point of view, some of the processes that lasers can induce in materials.

The field of laser-material interactions is inherently multidisciplinary. Upon impact of a laser beam on a material, electromagnetic energy is converted first into electronic excitation and then into thermal, chemical and mechanical energy. In the whole process the molecular structure as well as the shape of the material are changed in various ways. Understanding this sequence of events requires knowledge from several branches of physics. A unified presentation of the subject, for the benefit of the materials researcher as well as the advanced student, is attempted here. In order to keep the book reasonably trim, I have focused on laser effects in solids such as thin films and technological materials. Related topics not covered are laser-induced chemical reactions in gases and liquids and laser effects in organic or biological materials.

This monograph grew out of a series of lectures which I gave for graduate-level students in applied physics at the University of Bern. Its layout reflects the diversity of the subject - Chapter 2 draws essentially from physical optics, Chapters 3 and 4 from materials science, and Chapter 5 from fluid dynamics and plasma physics. While experts in certain fields covered here may find the treatment of their speciality rather less than exhaustive, I hope that the integrated treatment attempted will serve at least two purposes: It should provide access and orientation for students and newcomers in the large and diverse field of laser-material interactions, and it may, perhaps, uncover certain relations and connections not always obvious to the specialist. Readers' comments pointing out errors or inconsistencies will, in any case, be most welcome.

It is a pleasant duty to thank all those who have contributed to this book: colleagues who have given me access to and education on their work, as well as critical readers of preliminary versions of the manuscript who helped in eliminating some of its shortcomings. Finally, this book would not have appeared without the dedicated work of R. Flück and E. Krähenbühl who made most of the drawings and kept track of the paperwork, H.P. Weber who gave support, and - most important of all - my wife who smoothed the atmosphere and brewed many, many cups of coffee.

Bern, August 1986   *M. von Allmen*

Contents

1. Introduction

Laser beams, in a way, are a material scientist's dream: They can deliver concentrated "pure" energy to almost any material, and they do so exactly where, when, and in the quantity desired. Once available, lasers simply had to be used in materials science, if only because they were there. Things have, of course, turned out to be slightly more complicated, as we discuss in the following. Nevertheless, the fascination of transforming materials by laser beams will certainly continue to catch the fantasy of scientists and engineers for many years to come.

The historical development of the field has strongly been influenced by that of the laser technology itself. Experimental investigations of laser effects on materials began to appear soon after the first demonstration of the ruby laser in 1960. Due to the poor reproducibility and beam quality of the early lasers, this work was largely qualitative. It was mainly devoted to material evaporation, in which careful dosage of the laser energy is not so crucial. The popularity of the ruby laser was soon challenged by the newly developed Nd lasers (Nd-doped YAG or glass), which are inherently more stable and deliver optically superior beams, although they happen to emit in the near-infrared rather than the visible spectrum, which makes absorption in many materials more difficult. Later, with lasers and other equipment becoming more sophisticated, experiments on controllable material melting (without evaporation) became feasible, and laser processing of semiconductor structures and thin films began to attract enormous interest. Related activities have largely dominated the literature since the late seventies [1.1].

Nd lasers, unlike the ruby laser, are also capable of Continuous-Wave (CW) operation and offer the possibility of continuous, rather than pulsed, processing [1.2]. Steady progress is being made in the development of diode- rather than flashlamp-pumped Nd lasers which have dramatically improved energy efficiencies. For most of the rugged industrial applications like welding or cutting, however, the laser of choice is the CO_2 laser, because it is technologically simple; it compensates for its unfavorable infrared wavelength with versatility, high continuous power and energy-efficiency. Another CW laser often used is the Ar laser which emits in the green spectral regime, but it has a low efficiency and is not available for high power levels. A most promising new type is the excimer laser, a pulsed device which emits in the ultraviolet and combines high power with acceptable efficiency [1.4]. As they are becoming cheaper and more versatile,

1

excimer lasers are increasingly replacing some of the earlier laser types. With photon energies of $4 \div 7$ eV,[1] excimer beams are not only absorbed efficiently in most materials, but are also able to break chemical bonds directly upon absorption, i.e. before even heat is created. This opens up regimes of interaction not accessible with other laser sources.

An alternative to laser beams as energy sources in materials processing are particle beams, in particular medium-energy electron beams. They have certain advantages with respect to absorption [1.5], and machines capable of delivering pulse energies or CW powers adequate for efficient processing are becoming available. If the existing differences in absorption characteristics are taken into account, one usually finds that electron-beam-induced material effects are practically identical to effects due to laser irradiation. An obvious disadvantage of particle beams, apart from their more expensive hardware, is the necessity to perform the processing in vacuum.

1.1 Experimental Aspects

Performing experiments on laser-induced material effects requires (in addition to diagnostic elements) three basic pieces of equipment: A laser, a beam delivery system, and a fixture for the sample or workpiece. Laser types suitable for materials processing, and more specific equipment data are found in specialized periodicals [1.6]. A typical laboratory-type beam delivery system may consist of nothing more than a simple lens and a set of attenuating filters to adjust the beam power, while industrial systems often contain elaborate setups of adjustable mirrors and lenses capable of guiding the beam to any desired location within the working area. Optical fibres provide a convenient and flexible alternative for beam guiding, but low-loss fibres for CO_2 and excimer wavelengths are still being sought. In high-power applications the delivery system is usually fitted with a device protecting the beam impingement area, such as a screen against debris or an inert-gas nozzle. Adequate safety precautions should, in any case, be taken to protect operators from exposure to the laser beam and from debris [1.7].

The sample fixture in laboratory experiments typically consists of an adjustable table, sometimes fitted with automated scanning equipment, sometimes with a microscope. Scanning is usually done linearly by means of an x,y table, or alternatively in a circular way by some sort of turntable. Certain experiments (e.g., those involving reactive materials) must be performed in vacuum or in an inert atmosphere, and therefore require the sample fixture to be surrounded by a suitable chamber into which the beam is directed through optical windows.

---

[1] The symbol $\div$ is used throughout the text as a shorthand for "from - to" or "between".

Another important consideration in designing experiments on laser-solid interaction is diagnostics. For quantitative work, the beam power or pulse energy should be monitored by a suitable calibrated detector. Much of what is known today about the mechanism of absorption of intense light has been learned by monitoring (in real-time) the laser light reflected or transmitted by the sample. Other diagnostic techniques often employed in-situ during laser processing include spectroscopy of emitted light (including pyrometry), transient sample temperature or conductivity measurements, measurements of thermionic emission as well as acoustic measurements. Somewhat more costly techniques are real-time mass spectrometry of sample disintegration products, in-situ Auger or Rutherford backscattering spectroscopy and time-resolved X-ray diffraction. The real-time information gathered by this instrumentation is usually supplemented, of course, by subsequent examination of the irradiated material, involving all the analytical techniques currently available to the materials scientist.

1.2 Outline

This monograph treats laser effects in metals, semiconductors and insulators, with an emphasis on technological materials and structures. It is not, however, intended to be an instruction manual (although some technical information on specific processes is given), but aims rather at providing the physical insights from which instruction manuals are derived. Related topics such as laser-induced chemical reactions in gases and liquids [1.8], and laser effects in organic or biological materials [1.9], are beyond the scope of this text.

The organization of the book is as follows. After this introduction, Chap. 2 deals with the absorption of light in solids. Intense laser beams tend to modify the optical properties of the irradiated matter by a number of phenomena, and the bulk of the chapter is devoted to such effects - from simple heating through the production of free carriers to nonlinear-index phenomena. Also treated are coupling effects due to macroscopic phenomena like melting and surface rippling.

The remaining chapters describe phenomena due to absorbed laser light, loosely following a sequence of increasing sample temperature or energy density. Chapter 3 is concerned with solid-state thermal processing, usually done with continuous beams of moderate power. Here a section is devoted to analytical calculations of the sample temperature under various conditions of irradiation and material response, followed by discussions of various processes of current interest. Chapter 4 treats laser remelting in its various forms. To emphasize interconnections between the various phenomena observed in this regime, a section on modelling (including numerical simulation) and fundamental aspects of melting, soldification and phase formation precede the sections that describe individual applications. Finally, Chap.5 deals with the most "energetic" laser effects: evaporation and ionization. It starts out again with a section on fundamentals, mainly intended to facilitate understanding of the following sections, which then describe phenomena pertinent to a wide range of laser intensities - from drilling and welding to the production of hot vapors and plasmas. We end our survey by returning to materials and consider deposition of thin films from the laser-generated vapors.

2. Absorption of Laser Light

Laser light, in order to cause any lasting effect on a material, must first be absorbed. As trivial as this may sound, absorption very often turns out to be the most critical and cumbersome step in laser processing. An enormous amount of work has been dedicated to investigating laser absorption mechanisms under various circumstances, and a great deal can be learned from this work for the benefit of laser materials processing.

The absorption process can be thought of as a secondary "source" of energy inside the material. Whilst driven by the incident beam, it tends to develop its own dynamics and can behave in ways deviating from the laws of ordinary optics. It is this "secondary" source, rather than the beam emitted by the laser device, which determines what happens to the irradiated material.

Section 2.1 is meant as a refresher and as a basis for subsequent discussions. It summarizes the familiar optical properties of condensed matter, as far as they are relevant to absorption. The following two sections then treat modes of optical behavior influenced by intense laser irradiation, arising from atomistic and from macroscopic material responses, respectively.

2.1 Fundamental Optical Properties

2.1.1 Plane-Wave Propagation

The simplest form of light[1] is a monochromatic, linearly polarized plane wave. This will, for most of our purposes, be a sufficient approximation of a real laser beam. The electric field of a wave propagating in a homogeneous and nonabsorbing medium can be represented as

$$\mathbf{E} = \mathbf{E}_0 e^{i(2\pi z/\lambda - \omega t)} \tag{2.1}$$

where z is the coordinate along the direction of propagation, ω is the angular frequency, and λ is the wavelength. The last two quantities are related

---

[1] Here and in what follows, we shall interpret the words "optical" and "light" generously such as to cover the whole range of wavelengths (between roughly 0.1 and 10 μm) currently of interest in laser materials processing.

through the phase velocity c/n_1, c being the speed of light, and n_1 the refractive index of the medium [$n_1 = 1$ in vacuum; it is hardly different in air at standard temperature and pressure (stp)] by

$$\lambda = \frac{2\pi}{\omega} \frac{c}{n_1} \ .$$ (2.2)

An expression analogous to (2.1) also holds for the magnetic field **H**. The magnetic and electric field amplitudes are related by

$$\mathbf{H}_0 = n_1 \epsilon_0 c \mathbf{E}_0 \ ,$$ (2.3)

with ϵ_0 being the dielectric constant in vacuum. On average, the electric and magnetic fields each carry the same amount of energy. However, in the force **f** exerted by the electromagnetic wave on an electron

$$\mathbf{f} = -e \left[\mathbf{E} + (n_1/c)(\mathbf{v} \times \mathbf{H}) \right]$$ (2.4)

the contribution due to the magnetic field is smaller than that due to the electric field by a factor of the order of v/c (v being the electron velocity), and hence it is usually negligible. It is the term $-e\mathbf{E}$ in (2.4) that ultimately produces just about every phenomenon discussed in this book.

The energy flux per unit area of the wave is termed irradiance[2] and given by

$$I = \left| \mathbf{E} \times \mathbf{H} \right| = n_1 \epsilon_0 c \mathbf{E}_0^2 \ .$$ (2.5)

In the language of quantum mechanics, a wave of angular frequency ω and irradiance I corresponds to the flux $I/\hbar\omega$ of photons of energy $\hbar\omega$. Light interacts with matter only in portions of whole quanta. However, while quantum mechanics is required to understand the microscopic aspects of this process, the photon fluxes in intense laser beams are enormous and classical concepts are generally adequate to describe beam-solid interaction phenomena.

The concept of a beam implies that the irradiance is maximum near the optical axis and falls off laterally. The most common lateral distribution is a cylindrically symmetric Gaussian

$$I(r) = I_0 e^{-r^2/w^2}$$ (2.6)

---

[2] The term "intensity", often used instead, denotes the energy flux per unit solid angle.

where I_0 is the irradiance on axis ($r = 0$), and w is referred to as the beam radius. The total power of the beam is then

$$P = \pi w^2 I_0 .$$ (2.7)

The distribution (2.6) strictly applies only to a laser operated in its fundamental (TEM$_{00}$) resonator mode, but we shall use it as an approximation throughout. The wave front of a freely propagating Gaussian beam can usually be assumed to be approximately planar. In a first approximation, where diffraction effects are ignored, this holds also for a focused beam in the vicinity of the focal point [2.1].

In absorbing media the real refractive index n_1 must be replaced by a complex index $\mathbf{n} \equiv n_1 + in_2$. The meaning of n_2 (also called the *extinction coefficient*) becomes apparent when (2.2) is modified accordingly and inserted into (2.1): The electric field, upon propagation over a distance z, decreases by the factor $\exp(\omega n_2 z/c)$, indicating that some of the light energy is absorbed. The absorption coefficient for the irradiance (2.5) is

$$\alpha = -\frac{1}{I}\frac{dI}{dz} = \frac{2\omega n_2}{c} = \frac{4\pi n_2}{\lambda} .$$ (2.8)

The inverse of α is referred to as the *absorption length*.

In inhomogeneous media the refractive index varies in space - in certain crystals it even depends on the direction of propagation. Spatial variations of the refractive index deform the wave front, bend the beam path and cause secondary waves to split off from the primary one, as discussed in standard textbooks on optics. We only cite the well-known formula for reflection of a wave perpendicularly incident from vacuum or air onto the plane boundary of a solid with refractive index $\mathbf{n}$. The ratio R of reflected-to-incident irradiance is, in this case, given by the Fresnel expression

$$R = \left| \frac{\mathbf{n}-1}{\mathbf{n}+1} \right|^2 .$$ (2.9)

The reflectance and the absorption coefficient determine the amount of beam power absorbed within the material. The power density deposited at the depth z by a perpendicularly incident beam of irradiance I is

$$J_a(z) = I(1-R)\alpha\left\{ 1 - \exp\left[-\int_0^z \alpha(z')dz' \right] \right\} .$$ (2.10)

This expression represents our "secondary source". The integral in the exponential function is referred to as the optical thickness of the material between 0 and z. In opaque materials ($z \gg 1/\alpha$) the fraction of energy absorbed is determined by the quantity $(1-R)$ alone, also known as the *absorptance*.

2.1.2 Macroscopic Material Properties

The connection between the refractive index and the properties of the medium of propagation is formally provided by Maxwell's equations. For the case of a nonmagnetic and isotropic material of dielectric constant ϵ and conductivity σ they may be written as

$$\nabla \cdot \mathbf{E} = 0 , \tag{2.11}$$

$$\nabla \cdot \mathbf{H} = 0 , \tag{2.12}$$

$$\nabla \times \mathbf{E} = -(1/\epsilon_0 c^2)\partial \mathbf{H}/\partial t , \tag{2.13}$$

$$\nabla \times \mathbf{H} = \epsilon\epsilon_0 \partial \mathbf{E}/\partial t + \sigma \mathbf{E} . \tag{2.14}$$

Taking the curl of (2.13), recognizing that $\nabla \times (\nabla \times \mathbf{E}) \equiv \nabla (\nabla \cdot \mathbf{E}) - \nabla^2 \mathbf{E}$ and using (2.11, 14) yields the wave equation

$$\nabla^2 \mathbf{E} = \frac{\epsilon}{c^2} \frac{\partial^2 \mathbf{E}}{\partial t^2} + \frac{\sigma}{\epsilon_0 c^2} \frac{\partial \mathbf{E}}{\partial t} . \tag{2.15}$$

Insertion of the plane wave ansatz (2.1) and use of (2.2) with the complex index shows that the latter is related to ϵ and σ by

$$\mathbf{n}^2 = \epsilon + i\sigma/\epsilon_0 \omega \equiv \epsilon = \epsilon_1 + i\epsilon_2 . \tag{2.16}$$

The quantity ϵ defined by the right-hand side of (2.16) is the *complex dielectric function* which can be regarded as a generalized response function of the material. The real and imaginary parts of $\mathbf{n}$ and ϵ are related by

$$\epsilon_1 = n_1^2 - n_2^2 ; \quad \epsilon_2 = 2n_1 n_2 , \tag{2.17}$$

$$n_1^2 = \frac{|\epsilon| + \epsilon_1}{2} ; \quad n_2^2 = \frac{|\epsilon| - \epsilon_1}{2} . \tag{2.18}$$

The dielectric function fully describes the response of a material to weak electromagnetic irradiation. It depends on the light frequency in a manner determined by the microscopic structure of the material. Let us briefly review the main features of nonmetals (insulators or semiconductors) and metals.

2.1.3 Nonmetals

Insulators and semiconductors in the absence of excitation have only bound electrons and are basically transparent, except in the vicinity of what can be thought of as resonances. In the classical Lorentz model the electron is represented as a harmonic oscillator driven by the oscillating force of the wave's electric field (2.4). The oscillating electrons produce a macroscopic polarization of the material which superposes on the electric field of the incident wave. The dielectric function is simply the ratio of the total field (wave plus polarization) to the field of the wave alone. In quantum mechanics, a resonance corresponds to the transition of an electron between two states, the energy difference ΔE of which determines the resonance frequency $\omega_0 = \Delta E / \hbar$. The classical and quantum-mechanical treatments yield almost identical expressions for the dielectric function. For a nonmetal with N_e bound electrons showing one single resonance we have [2.2]

$$\epsilon = 1 + \frac{N_e e^2}{m_e \epsilon_0} f_{osc} \frac{\omega^2 - \omega_0^2 + i\Gamma\omega}{(\omega^2 - \omega_0^2)^2 - \Gamma^2 \omega^2} . \tag{2.19}$$

Here f_{osc}, the *oscillator strength*, is a measure of the probability of the transition, while the *damping constant* Γ describes the width of the resonance that arises from the finite width of the initial and final electron states. The optical consequences of a resonance are illustrated in Fig.2.1. It shows ϵ, along with the derived quantities n, R and α, as a function of the light frequency, as obtained from (2.8, 9, 17-19). Note that a peak in the absorption coefficient due to a resonance is invariably accompanied by a peak in the reflectance.

Real materials have, of course, many resonances, and in their expression for ϵ the resonance term of (2.19) is replaced by a sum over many terms. The most important resonance arises from transitions of valence-band electrons to the conduction band (interband transitions). To induce an interband transition the incident photon must have an energy at least equal to the band-gap energy E_g. The free carriers (electrons and holes) created in pairs in interband transitions can, if present in sufficient numbers, influence the optical response of the material. Insulators have band gaps corresponding to light frequencies in the vacuum ultraviolet, and carrier concentrations under day-light illumination remain negligible. Semiconductors, on the other hand, have band gaps in the visible or infrared part of the spectrum. Free carriers (optically or thermally generated) contribute measurably to the metal-like reflectance of many semiconductors in the visible region.

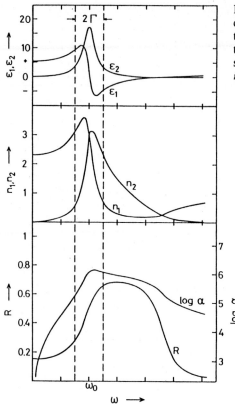

Fig.2.1. Frequency dependence of the dielectric function, the refractive index, the Fresnel reflectance and the absorption coefficient for a medium with a single resonance at ω_0 (calculated for $\hbar\omega_0 = 4\,\text{eV}$, $\hbar\Gamma = 1\,\text{eV}$, $N = 5\cdot 10^{22}\,\text{cm}^{-3}$)

Most nonmetals show - in addition to electronic transitions - resonant coupling to high-frequency optical phonons located in the near-infrared region of the spectrum. Phonon coupling can be described by resonance terms similar to the one in (2.19), but with masses and damping constants characteristic of lattice vibrations rather than electronic vibrations.

Figures 2.2,3 show the absorption coefficient and the reflectance as a function of wavelength for two representative nonmetals - quartz (E_g = 6.9eV, corresponding to a wavelength of 180nm) and crystalline Si (E_g = 1.1eV at room temperature, corresponding to $\lambda = 1.13\,\mu\text{m}$). Both materials exhibit the same essential features - absorption peaks around 10 μm due to phonon coupling, weak absorption at intermediate wavelengths and a steep increase in absorption as the photon energy approaches the band gap.

With reference to Fig.2.3, it will be noted that the absorption coefficient of Si increases with decreasing laser wavelength only gradually near 1.13 μm ($\hbar\omega = 1.1\,\text{eV}$), but rather abruptly near 0.36 μm ($\hbar\omega = 3.4\,\text{eV}$). The explanation is that there are two different kinds of interband transitions in materials like Si. The first kind, direct transitions, lead from valence-band to conduction-band states with the same wave vector. Such transitions are

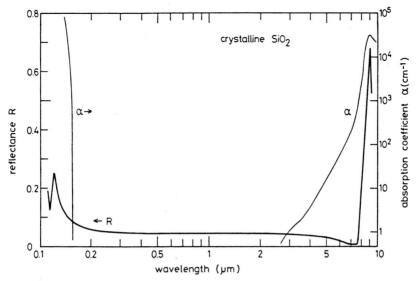

Fig.2.2. Reflectance and absorption coefficient as a function of wavelength for crystalline quartz

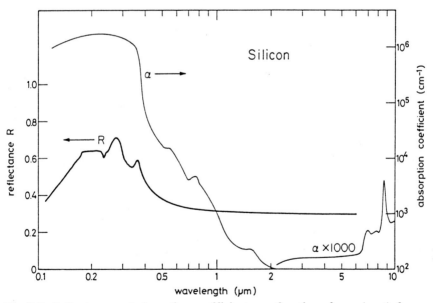

Fig.2.3. Reflectance and absorption coefficient as a function of wavelength for crystalline silicon

possible in Si only at photon energies exceeding its direct gap at 3.4 eV. The second kind of interband transitions, indirect transitions, involve valence- and conduction-band states of different wave vectors. The difference in wave vector must be provided by a phonon, as the photon itself carries only negligible momentum. Since the fundamental gap at 1.1 eV is an indirect gap, it allows only phonon-assisted transitions. The probability of such transitions depends on the phonon occupancy and is relatively small and temperature-dependent. Other indirect-gap materials are Ge, diamond, SiC, GaP, AlP and AlAs, whereas materials such as Se, Te, ZnO, as well as most II-VI and IV-VI compounds have direct fundamental gaps.

The simple picture outlined above applies only to homogeneous materials. In nonmetals inhomogeneous on a scale of one wavelength or more the optical appearance is modified by light scattering at grain boundaries or inclusions. This results in quite strong absorption even in materials that, intrinsically, would be transparent, e.g., ceramics. The effect can be rationalized in terms of an effective multiplication of the light path inside the material caused by a large number of scattering events. Inclusions that are small compared to the wavelength can, on the other hand, be treated in terms of an "averaged" dielectric function. An expression often used to describe randomly distributed inclusions (nonmetallic or metallic) with the dielectric function ϵ_i in a matrix with the dielectric function ϵ_m was first derived by Maxwell-Garnett in order to explain the colors shown by certain suspensions or "colored glasses". The average dielectric function ϵ follows from [2.4]

$$\frac{\epsilon - \epsilon_m}{\epsilon + 2\epsilon_m} = \frac{\Xi(\epsilon_i - \epsilon_m)}{\epsilon_i + 2\epsilon_m} \tag{2.20}$$

where $\Xi < 1$ denotes the volume fraction occupied by the inclusions. Eq. (2.20) predicts "resonances" leading to absorption peaks that are present neither in the pure host nor in the inclusion material.

2.1.4 Metals

The optical response of a metal is dominated by the conduction electrons. Since the electron gas is degenerate, only electrons in states close to the Fermi level, referred to as *free electrons*, contribute to the optical properties. There is no resonance frequency for a free electron, and its only interaction with the lattice is by collisions. The dielectric function of a free-electron metal can be obtained from (2.19) by formally replacing the damping constant Γ by the inverse collision time $1/\tau_e$ and setting the resonance frequency equal to zero, and f_{osc} equal to one. The resulting expression can be written as

$$\epsilon = 1 + \omega_p{}^2 \frac{-\tau_e{}^2 + i\tau_e/\omega}{1 + \omega^2 \tau_e{}^2} \tag{2.21}$$

where

$$\omega_p = \sqrt{\frac{N_e e^2}{m_e \epsilon_0}} \tag{2.22}$$

is the electron *plasma frequency*. At $\omega = \omega_p$ (which is in the vacuum ultra-violet for most metals) both ϵ_1 and n_1 vanish. The variation of ϵ and the associated quantities with the light frequency is shown in Fig.2.4. The plasma frequency is seen to separate two regimes of rather different optical properties: Large R and α for $\omega < \omega_p$, and small R and α for $\omega > \omega_p$.

The optical properties of a free-electron metal for $\omega < \omega_p$ are related to its DC conductivity σ_0. With the aid of the Drude formula [2.2, 3]

$$\sigma_0 = N_e e^2 \frac{\tau_e}{m_e} = \omega_p{}^2 \epsilon_0 \tau_e , \tag{2.23}$$

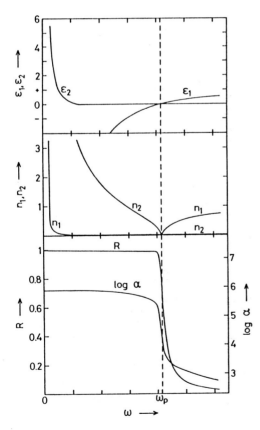

Fig.2.4. Frequency dependence of the dielectric function, the refractive index, the Fresnel reflectance and the absorption coefficient of a free-electron metal (calculated for $\hbar\omega_p = 8.3\,\text{eV}$, corresponding to $N_e = 5 \cdot 10^{22}\,\text{cm}^{-3}$, and $\hbar/\tau_e = 0.02\,\text{eV}$)

useful approximations for the optical parameters can be obtained. For the range $\omega \ll 1/\tau_e$ (far-infrared region) Eq.(2.21) gives $\epsilon_1 \simeq -\sigma_0 \tau_e/\epsilon_0$ and $\epsilon_2 \simeq \sigma_0/\epsilon_0 \omega$, from which follows that $n_1 \simeq n_2 \simeq (\sigma_0/2\epsilon_0 \omega)^{1/2}$. Using (2.8, 9) yields

$$1 - R \simeq \sqrt{8\epsilon_0 \omega/\sigma_0} \tag{2.24}$$

and

$$\alpha \simeq \sqrt{2\omega\sigma_0/\epsilon_0 c^2} . \tag{2.25}$$

For the range $1/\tau_e < \omega < \omega_p$ (near-infrared and visible regions for most metals) we have $n_1 \simeq \omega_p/2\omega^2 \tau_e \simeq 0$ and $n_2 \simeq \omega_p/\omega$ which gives

$$1 - R \simeq \frac{2}{\omega_p \tau_e} = \frac{2\epsilon_0 \omega_p}{\sigma_0} \tag{2.26}$$

and

$$\alpha \simeq \frac{2\omega_p}{c} . \tag{2.27}$$

In real metals the simple free-electron behavior is modified by a number of secondary effects, in particular by interband transitions. As an illustration, we depict in Fig.2.5 the coupling parameters for Al and Au.

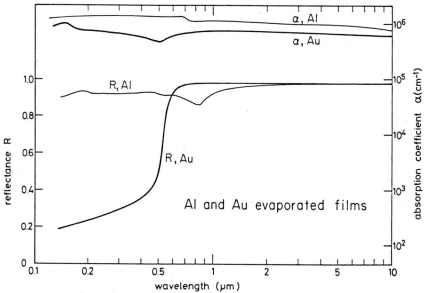

Fig.2.5. Reflectance and absorption coefficient as a function of wavelength for aluminum and gold

The curves of both metals show oscillations caused by interband transitons, such as the one at 0.83 μm (1.5eV) in Al. A comparison with Fig.2.4 indicates that the plasma frequency for Al is indeed in the vacuum ultraviolet, whereas that of Au appears to be in the visible. The anomalously low apparent plasma frequency of Au (which causes its yellow color) is due to transitions from d-band states, which set in at $\hbar\omega \simeq 2$ eV. The shift in ω_p is readily understood from an inspection of Figs.2.1,4: The d electrons make a bound-type contribution to the dielectric function, and since the contribution to ϵ_1 is positive, the point $\epsilon_1 = 0$ which defines ω_p is shifted to a lower frequency. Similar behavior is also found in Cu and Ag.

The optical properties predicted by Fig.2.4 are, of course, bulk properties. In practice, metal surfaces usually show lower reflectance than the bulk due to contamination (adsorbates, oxide layers, etc.) or macroscopic defects. This explains the appreciable scatter in the reflectivity data found in the literature. Deviations from bulk behavior may also arise from "intrinsic" surface phenomena such as plasmon excitation or diffuse electron scattering, particularly in thin films. The optical behavior becomes almost completely dominated by surface effects when the dimension of the metal is shrunk to values of the order of the absorption length, such as in extremely thin evaporated films or in aggregate structures consisting of small insulated metallic particles [2.4].

This concludes our short survey of the "linear" optical properties of materials. Some experimental values of reflectance and absorption length for a number of nonmetals and metals are given in Table A.1.

2.2 Modified Optical Properties

The fundamental optical properties outlined in the previous section represent the response of materials to light that is weak enough so as not to perturb the states of the electrons and atoms significantly. Powerful laser irradiation, however, is known to alter the optical properties of many materials, and often drastically so. Coupling is then no longer well characterized by a static dielectric function but becomes a dynamical process. The amount of light absorbed in the material may turn out to be smaller or larger than expected from the fundamental optical properties.

All beam-induced changes in the optical properties of solids can be traced back to one of three mechanisms. These are, roughly in a sequence of increasing irradiance, the following:

(i) Heat production and resulting changes in the density or the electronic characteristics of the material. Pertinent effects include thermal self-focusing in transparent media as well as "thermal runaway" phenomena in semiconductors and metals.

(ii) Optical generation of free carriers by interband transitions or impact ionization in semiconductors and insulators. As a result, the absorption coefficient increases dramatically, possibly causing explosive material damage.

(iii) Nonlinear distortion of electron orbitals or whole molecules by the electric field of an intense beam. A host of nonlinear optical phenomena, including self-focusing and multiphoton absorption, are caused by field effects.

Apart from these "intrinsic" phenomena, beam-solid coupling, in addition, tends to be affected by beam-induced changes in the *shape* of the material, usually in connection with melting or evaporation. These will be discussed in Sect.2.3. In what follows we shall consider phenomena in solids as related to the three mechanisms mentioned above.

2.2.1 Self-Focusing

Self-Focusing (SF) - or its opposite, self-defocusing - occurs if the real part of the refractive index of the medium varies locally as a function of irradiance. Extensive reviews on this topic were presented by *Akhmanov* [2.5], *Svelto* [2.6], and *Shen* [2.7]. Since laser beams are more intense near the beam axis than away from it, an irradiance-dependent real index of refraction has an effect somewhat similar to a lens placed in the beam path. The "lens", as illustrated in Fig.2.6, is converging if n_1 increases with the irradiance - the phase velocity decreases towards the beam axis, causing a plane wave front to become concave and eventually to collapse. Similarly, the "lens" is diverging if n_1 decreases as a function of irradiance. This self-de-

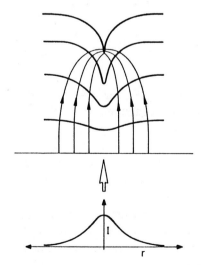

Fig.2.6. Geometrical rays (arrows) and phase fronts of a Gaussian beam undergoing self-focusing in a medium in which the local real refractive index increases as a function of irradiance

focussing is inconsequential as far as energy deposition is concerned (although it can be disturbing), but self-focusing increases the effective irradiance and reinforces other irradiance-dependent phenomena. The variation in refractive index can itself be thermal or field induced.

Let us first consider *thermal self-focusing*. The temperature dependence of the refractive index can be thought of as consisting of two parts

$$\frac{dn_1}{dT} = \left(\frac{\partial n_1}{\partial T}\right)_\rho + \left(\frac{\partial n_1}{\partial \rho}\right) \frac{d\rho}{dT} . \tag{2.28}$$

The first term accounts for the temperature dependence of the electronic or molecular polarizability due to shifts in the absorption bands. Thus, in materials with band-gap energies in the ultraviolet region that decrease with temperature, this term tends to be positive in the visible spectrum. The second term, connected to thermal expansion (ρ being the density of the solid), is usually negative since $\partial n_1/\partial \rho > 0$ and $d\rho/dT < 0$. (In addition, there may be strain-induced birefringence, which we do not consider here). Hence, the first term may favor self-focusing while the second tends to oppose it. However, the two contributions to (2.28) have different relaxation times (thermal expansion proceeds at the sound velocity and is much "slower" than heating itself), and there may be transient self-focusing for a short pulse even if the steady-state value of dn_1/dT is negative.

To describe thermal SF the equations for beam propagation have to be solved simultaneously with the associated heat-conduction problem [2.5]. Consider a weakly absorbing medium in which $dn_1/dT > 0$, irradiated by a parallel Gaussian beam of radius w and power P. The absorbed power creates a continuous positive thermal lens in the material along the beam path. The minimum absorbed power required for the thermal lens to balance beam diffusion by diffraction in steady state is found to be

$$P_{tr} = \frac{\lambda K}{dn_1/dT} \tag{2.29}$$

where K is the thermal conductivity. For P exceeding P_{tr} the beam focusses after propagating the distance (aberrationless approximation)

$$z_f \simeq \sqrt{\pi z_0 \frac{P_{tr}}{\alpha P}} \tag{2.30}$$

where $z_0 = 2\pi w^2 n_1/\lambda$ is the diffraction length of the beam. In many materials thermal SF may take place at powers as little as 1 W. Under certain conditions the focused beam can produce a thermal waveguide in the medium where it travels for distances $\gg z_0$ without divergence. In a typical experiment [2.8], a self-guided filament about 50 μm in diameter, resulting from thermal SF of a 3 W Ar laser beam in lead glass ($n_1^{(0)} = 1.75$, $dn_1/dT \simeq 10^{-5}K^{-1}$, $K \simeq 5.4 \cdot 10^{-3}W/cm \cdot K$ and $\alpha \simeq 0.1cm^{-1}$), was observed at a distance of 20 cm.

Electric-field-induced SF, unlike its thermal counterpart, is a high-irradiance effect relevant only for powerful nano- or picosecond pulses. The root of field-induced changes in the refractive index is the anharmonicity of all interparticle potentials. As the force (2.4) grows stronger, the linear relation between the electric field and the polarization is eventually lost and the harmonic oscillator model, on which (2.19) is based, fails to reproduce the interaction accurately. The true motion of the oscillating electrons now contains Fourier components at frequencies other than the driving frequency: the material polarization, and hence the dielectric function, show resonances at multiples of ω (including at $\omega = 0$). Related physical phenomena include the generation of frequency-shifted secondary waves (frequency multiplication, optical mixing and stimulated scattering), as well as a field-dependent dielectric function for the incident wave. The latter is commonly written as

$$\epsilon = \epsilon^{(0)} + \epsilon^{(2)} E^2 + \epsilon^{(4)} E^4 + ... \tag{2.31}$$

where $\epsilon^{(0)}$ is the linear dielectric function. Electronic as well as molecular polarizability contributes to the nonlinear terms in (2.31). In isotropic media a major nonlinear contribution is due to electrostriction. In liquids strong fields are known to induce orientation of molecules, making the medium optically anisotropic and increasing the average value of ϵ_1 (high-frequency Kerr effect). The nonlinear parameters $\epsilon^{(n)}$ in (2.31) are, in general, complex. For the present discussion we are interested in the real parts; the imaginary parts are related to multiphoton absorption (Sect.2.2.4).

In dealing with field-induced SF it is convenient to introduce an intensity-dependent real refractive index of the form

$$n_1 = n_1^{(0)} + \gamma I \tag{2.32}$$

where $n_1^{(0)}$ is the linear index and $\gamma = \epsilon^{(2)}/2c\epsilon_0\epsilon^{(0)}$.

Consider a transparent medium with $\gamma > 0$. In the geometrical-optics approximation the effect of SF on a Gaussian beam can be described in terms of a field-induced thin lens with a focal length that depends on irradiance [2.6]

$$z_{nl} \simeq w \sqrt{2n_1^{(0)}/\gamma I} . \tag{2.33}$$

The lens exactly balances diffraction if z_{nl} equals z_0. This holds, independently of the incident beam radius, if the power is equal to the threshold value

$$P_{tr} \simeq \frac{\lambda^2}{2\pi\gamma n_1^{(0)}} . \qquad (2.34)$$

In the optical glass "BK-7", γ is $4 \cdot 10^{-16}$ cm^2/W, for which (2.34) predicts a threshold of about 3 MW for green light (allowance for aberrations leads to a slightly different numerical factor). The situation in which SF and diffraction just balance is known as self-trapping. For $P > P_{tr}$ the beam collapses into a focal spot after propagating along the distance (aberrationless approximation)

$$z_f = \frac{z_0}{\sqrt{P/P_{tr} - 1}} . \qquad (2.35)$$

As the beam power varies in time so does z_f and the point of self-focus moves back and forth along the optical axis. However, the above equations hold only for pulses long compared to the response time of the nonlinear process. For electronic polarizability this is of the order of 10^{-14} s, while coordinated molecular motion, as involved in the Kerr effect, takes of the order of 10^{-11} s. Electrostriction has a time constant given by the transit time of a sound wave across the beam diameter, i.e., 10^{-9} s or more (time constants for thermal SF can be as long as 1 s). Whenever the pulse duration is of the order of the relevant time constant, SF is of a transient nature. Qualitatively, the picture is that the index nonlinearity experienced by the leading edge of a pulse is smaller than that seen by the trailing edge. Accordingly, only the lagging portion of a pulse experiences SF. For short pulses relaxation effects may result in effective suppression of SF. Besides, SF is not limited to narrow Gaussian beams but can occur in beams of arbitrary diameter if their lateral profile is modulated. Individual peaks or crests of sufficient power can undergo SF independently, causing broad powerful beams to break up into many small filaments inside a nonlinear medium, e.g., a solid-state laser amplifier. Self-focusing has also been observed in gases [2.9].

The arguments considered so far would predict an infinite irradiance to result, for diffraction cannot prevent the beam from collapsing into a point-like focal spot once $P > P_{tr}$. In reality, terminal focal diameters are found to be at least a few μm. The mechanism ultimately limiting SF is the generation of free carriers, either by thermal emission in absorbing media or by optical breakdown in nonabsorbing ones [2.10]. The free-carrier contribution to the real part of the refractive index is negative (see below), and it will eventually cause self-defocusing. At the same time, the free carriers cause strong absorption. The typical result is explosive thermal damage in the focal region.

2.2.2 Phenomena Due to Free-Carrier Generation

Free-carrier generation is the most important self-induced coupling effect in nonmetals. Before discussing mechanisms of free-carrier generation let us briefly consider the impact of free carriers on the optical properties of a nonmetal.

The simplest approach is to think of the total material polarization, and hence of the dielectric function, as a sum of lattice and carrier contributions. Remembering (2.21, 22) and considering electrons only, we can write the total refractive index as (assuming $\omega \gg 1/\tau_e$)

$$\mathbf{n} = \mathbf{n}_0 \sqrt{1 + (\omega_p/\mathbf{n}_0\omega)^2(-1 + i/\omega\tau_e)} \qquad (2.36)$$

where $\mathbf{n}_0$ is the complex lattice index and $\omega_p = (N_e e^2/m_e \epsilon_0)^{1/2}$, with N_e being some function of the irradiance. It is assumed here that N_e is small compared to the atomic density and, in particular, small enough that the lattice and the carriers behave independently of each other (a condition often violated in practice, as we shall see later on). A term analogous to the electron term applies for holes. The effect of free carriers on the optical properties of the nonmetal, as obvious from (2.36), is to reduce the real part and to increase the imaginary part of $\mathbf{n}$. This, according to (2.8, 9), increases the absorption coefficient and tends to decrease the reflectance for $\omega_p < \omega$. Only for carrier densities large enough that $\omega_p > \omega$ does the reflectance increase by further increasing ω_p.

In semiconductors, holes tend to be mobile and contribute significantly to absorption. Since electrons and holes come in equal numbers, it is convenient to treat them together and to write the total absorption coefficient in the form

$$\alpha = \alpha_0 + N_{eh} \Sigma_{eh} \qquad (2.37)$$

where α_0 is the lattice absorption coefficient, $N_{eh} = N_e = N_h$ is the density of carrier pairs, and Σ_{eh} is the absorption cross section of a carrier pair, defined as

$$\Sigma_{eh} = \frac{e^2}{\epsilon_0 n_1 c\omega^2}\left(\frac{1}{m_e^* \tau_e} + \frac{1}{m_h^* \tau_h}\right). \qquad (2.38)$$

Here m_x^* and τ_x ($x = e, h$) are the effective masses and collision times of electrons and holes, respectively [2.11]. Note that Σ_{eh} scales essentially with λ^2, making free-carrier absorption mainly relevant for infrared beams. For example, in Si experimental values of Σ_{eh} at 20°C are around $5 \cdot 10^{-18}$ cm^{-2} at 1.06 μm [2.12] and around $1 \cdot 10^{-16}$ cm^{-2} at 10.6 μm [2.13].

Let us now consider mechanisms of carrier generation. Free-carrier phenomena are of a somewhat different nature in semiconductors and in insulators, and we shall consider the two types of material separately.

2.2.3 Semiconductors

Even far-infrared beams with $\hbar\omega < E_g$ and of moderate irradiance can create significant densities of free carriers in semiconductors, due to thermal excitation. The equilibrium free-carrier density in an intrinsic semiconductor at temperature T is given by

$$\bar{N}_{eh}(T) = 2\left(\frac{kT}{2\pi\hbar^2}\right)^{3/2} (m_e^* m_h^*)^{3/4} \exp\left(-\frac{E_g}{2kT}\right) \tag{2.39}$$

from which it follows that the absorption coefficient (2.37) increases with temperature faster than $\exp(1/T)$ if α_0 is constant; in applying (2.39) it is generally necessary to allow for a temperature dependence of the band gap [2.14]. For illustration, Fig.2.7 shows measured and calculated absorption coefficients at 10.6 μm in crystalline n-type Si of various doping levels as a function of temperature [2.15]. The additional absorption by the free carriers causes stronger heating which leads to even more carriers in a sort of positive feedback loop. This results in typical "thermal runaway" behavior, observed particularly with far-infrared radiation where lattice absorption is small and free carriers absorb efficiently.

Photons with $\hbar\omega > E_g$ create carriers by a mechanism far more efficient than heating, namely interband absorption, in which every absorbed photon leaves behind a carrier pair. Optically generated carriers and their effects have received considerable attention in connection with laser annealing of Si by nanosecond pulses. The rate of generation of carrier pairs by interband absorption in an intrinsic semiconductor is $\alpha_0 I/\hbar\omega$, were α_0 is the coefficient for interband absorption, and I is the local irradiance which depends on both interband and free-carrier absorption [2.16]. To model the situation, carrier generation as well as carrier losses, due to diffusion and recombination, must be allowed for. The resulting carrier density follows from a differential equation of the form

$$\frac{\partial}{\partial t} N_{eh} = \alpha_0 \frac{I}{\hbar\omega} + \nabla(D_{amb} \nabla N_{eh}) - \frac{1}{\tau_r}[N_{eh} - \bar{N}_{eh}(T)] \tag{2.40}$$

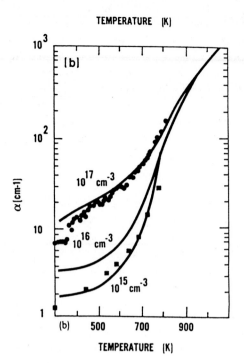

Fig.2.7. Absorption coefficient at $\lambda =$ 10.6 μm in phosphorus-doped Si at various doping levels, as a function of temperature. (Dots and squares are measured, solid lines calculated) [2.15]

where the first term describes generation and the second diffusion of carrier pairs. The ambipolar diffusion constant D_{amb} allows for the attraction between electrons and holes. The last term of (2.40) is the recombination rate (for $\overline{N}_{eh} > N_{eh}$ this term would describe thermal carrier generation). The recombination lifetime τ_r must be considered as a function of both temperature and carrier density. In most semiconductors the dominant recombination mode at large carrier densities is Auger recombination in which the recombination energy E_g is given to a third carrier (electron or hole), rather than to the lattice. Auger transitions are thus three-particle processes with $1/\tau_r \propto N_{eh}^2$.

Under conditions typical for pulsed laser annealing, the carrier density can reach values where the energy absorbed by the free carriers exceeds that absorbed by the lattice. Moreover, at densities above 10^{17} or 10^{18} cm^{-3} collisions among the carriers begin to dominate over collisions with the lattice, and carriers behave collectively, i.e., as a plasma. It is well known that thermally excited carriers in a semiconductor have a Boltzmann distribution - this is merely the limiting form of the Fermi distribution for carriers with $E-E_F \gg kT$. Optically excited carriers tend to have more complicated distributions, depending on the properties of the light source as well as the solid. Light absorption by free carriers sends electrons into states high up in the conduction band and holes deep down in the valence band. Carriers excited by short, intense pulses "remember" the pulse spectrum in their energy distribution [2.17]. This nonthermal distribution relaxes by means of carrier-lattice as well as carrier-carrier collisions. Whilst at low concentrations carriers reach thermal equilibrium with the lattice individually, in the plasma regime a thermal distribution is established among the carriers before equilibrium with the lattice is reached [2.18]. One may then speak of carrier (electron and hole) temperatures that differ from the lattice temperature. In a Boltzmann plasma the carrier temperature T_c is related to the average carrier excess energy $\langle E_c \rangle$ by $\langle E_c \rangle = (3/2)kT_c$ (c = e or h). Electron and hole plasmas exchange energy by e-h collisions and by Auger processes, and for times long enough for the electrons to thermalize among themselves electron and hole temperatures may be taken as equal. A schematic representation of the electron plasma inside a laser-irradiated semiconductor, indicating typical values for the various time constants involved, is shown in Fig.2.8.

Let us now consider the role played by the carrier plasma in the beam-solid interaction. The plasma may be regarded as an independent system which absorbs light and exchanges energy with the lattice. The plasma gains energy from interband absorption (an amount of $\hbar\omega - E_g$ per carrier pair) as well as from free-carrier absorption. Furthermore, if Auger processes dominate the recombination, an amount E_g is added to the plasma energy for every recombination event. The plasma loses energy by carrier diffusion - every carrier pair carrying away its mean excess energy $\langle E_c \rangle$ - and by phonon emission which heats the lattice. The latter is, apart from residual phonon absorption, the only source of lattice heating. The plasma thus acts as a sort of funnel through which the absorbed energy must go before it is

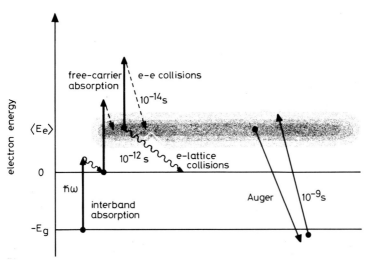

Fig.2.8. Schematic diagram of electron energies in a laser-irradiated semiconductor, showing the free-electron plasma exchanging energy with the beam and the lattice. An analogous diagram holds for holes

given to the lattice. The resulting average excess carrier energy - taking electrons and holes to be in equilibrium with each other - follows from an expression of the form [2.19]

$$N_{eh} \frac{\partial}{\partial t} \langle E_c \rangle = I[\alpha_0(1 - E_g/\hbar\omega) + N_{eh}\Sigma_{eh}] + (E_g/\tau_r)[N_{eh} - \bar{N}_{eh}(T)]$$

$$+ \nabla(\langle E_c \rangle D_{amb} \nabla N_{eh}) - (\langle E_c \rangle - 3kT/2)N_{eh}/\tau_E . \qquad (2.41)$$

The first term represents energy gained from interband and free-carrier absorption, while the second is the energy liberated upon Auger recombination. The last two terms represent energy lost to diffusing carriers and energy transferred to the lattice, respectively. Here $3kT/2$ is the equilibrium carrier excess energy, and τ_E is an effective carrier-lattice energy relaxation time. Eq.(2.41) must be solved together with (2.40) and the heat-flow equation for the lattice temperature. If the irradiance changes slowly in comparison to the time constants in (2.41), the plasma density and energy will assume steady-state values which make gain and loss rates equal. The main difficulty in such a calculation is that the material parameters D_{amb} E_g, α_0, Σ_{eh}, τ_r and τ_E depend on the lattice temperature and, via screening effects, also on the density and temperature of the plasma. In Si, carrier lifetimes and energy relaxation time are expected to saturate at carrier densities of the order of 10^{21} cm^3 [2.18]. As an illustration of this regime, Fig.2.9 exhibits the surface temperature of crystalline Si irradiated by a Nd laser pulse, as calculated numerically by *Lietoila* and *Gibbons* [2.19], with and without inclusion of absorption by the carrier plasma (also shown is the influence of the temperature dependence of E_g). At the instant when the lattice reaches the melting point (1412°C), the carrier density turns out to be $2 \cdot 10^{20}$ cm^{-3}.

The assumption of a Boltzmann distribution of carriers is justified only as long as the carrier density remains small compared to the density of states in the energy range covered by the plasma; otherwise the plasma becomes degenerate. The main effect of degeneracy is to limit the rate of energy exchange between carriers due to the lack of available unoccupied states. Also affected, and for the same reason, is the Auger recombination rate which is found to increase only as the square, instead of the cube, of the carrier density in degenerate plasmas [2.20]. The decreasing rates of carrier scattering and recombination cause an increase in the energy diffusing out of the light-absorbing zone by the flow of hot carriers. *Yoffa* [2.18] predicted that the ambipolar diffusivity D_{amb} should become larger than usual thermal diffusivities by about 3 orders of magnitude. Under such circumstances the spatial distribution of deposited energy would clearly be determined by the carrier diffusion length, rather than by the light-absorption length. On the other hand, diffusion may to some extent be counteracted by

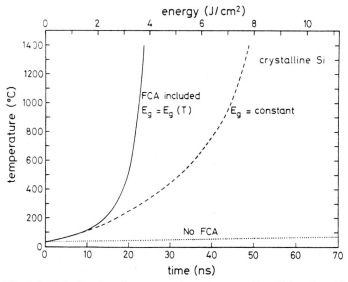

Fig.2.9. Calculated surface temperature in crystalline Si irradiated by a Nd-laser beam of 160 MW/cm² as a function of time (solid line). Different curves result if free carrier absorption (FCA) is ignored (dotted) or included with a constant (dashed), rather than temperature-dependent, band gap. [2.19]

a local decrease in band gap caused by lattice heating or by plasma effects, which tends to confine the carrier plasma (see below).

Dense carrier plasmas can store energy not only as thermal energy but also in the form of plasma oscillations. A share of the order of 10% of the total plasma energy was estimated to consist of plasmons in laser-irradiated Si with $N_c \simeq 10^{20}$ cm^{-3} [2.18]. Plasmons can be treated as quasi-particles which dissipate energy in collisions with the lattice. The interaction is strongest if the plasmon energy equals the relevant phonon energy (typically below 0.1 eV, corresponding to carrier densities below 10^{17} cm^{-3}). In much denser plasmas, plasmons tend to decouple from the lattice unless the plasmon energy reaches or exceeds E_g. In this event plasmon decay via e-h pair creation as well as recombination by plasmon emission become possible. Such processes have been observed in Ge and Si irradiated by picosecond pulses [2.21, 22].

The creation of a dense carrier plasma, which involves the removal of a significant fraction of the binding electrons, should have an impact also on the properties of the lattice itself. First, moving electrons from valence-band to conduction-band states reduces the size of the band gap in most semiconductors, while in some (like HgTe and HgSe) it causes the reverse effect [2.23]. A local narrowing of the band gap - be it caused by plasma effects or by local lattice heating - creates an electric field which tends to confine the plasma. *Lietoila* and *Gibbons* [2.19] included the effect in their computer modeling of laser-irradiated Si (Fig.2.9), but found it to be insignificant in their case.

The plasma should also affect the mechanical strength of the semiconductor lattice. The removal of electrons from (bonding) valence-band to (antibonding or plane wave) conduction-band states tends to soften the lattice and to reduce the phonon frequencies. The reduced phonon energies result in an increased number of excited phonons at any given temperature, further destabilizing the lattice. *Van Vechten* and associates [2.24] proposed that a large density of photoexcited carriers could lead to a fluid-like state of the material at lattice temperatures far below the normal melting point, thereby bringing about the observed crystallization of amorphous Si by nanosecond laser pulses without thermal melting. As main experimental support for this hypothesis the researchers cited measurements of lattice temperature based on Raman scattering, which indicate insignificant lattice heating even for irradiances above the observed annealing threshold [2.25]. The interpretation of the Raman data remains a subject of controversy, however [2.26]. Particularly hard to explain by the plasma annealing hypothesis is the observed duration (up to some 10^{-7}s) of a state of enhanced reflectivity of the irradiated Si [2.27], readily explained by the metallic nature of molten Si if a thermal melting mechanism is assumed. Deformation-potential scattering of energetic electrons ($E_e \simeq 1\,eV$) in Si yields energy relaxation times of the order of 10^{-12} s, and screening seems unable to increase this time by anything near the amount required [2.28]. Circumstantial experimental evidence for thermal melting in nanosecond laser annealing of Si has been presented by a number of researchers [2.29-31]. Experiments with picosecond and sub-picosecond pulses indicate that the relaxation of energetic carriers in Si is much faster than that of thermal ones. *Malvezzi* et al. [2.32] inferred ambipolar diffusivities of 1/20 of the near-equilibrium value, leading to a hot-carrier diffusion length of only some 10 nm. These and other experiments show that thermal melting in Si is delayed by plasma effects by at most a few picoseconds [2.33]. As an illustration, Fig.2.10 shows time-resolved reflectance measured at $\lambda = 1$ μm in crystalline Si irradiated by red 90 femtosecond pulses ($\lambda = 620\,nm$) of various fluences [2.34]. Here, for $0.63F_{th}$ (F_{th} denoting the threshold fluence for surface melting) the reflectance is seen to decrease, indicating that the free-carrier density corresponds to the condition $\omega_p < \omega$ for the probe frequency. The reflectance then slowly returns to its static value as the carrier plasma diffuses away from the surface. For fluences at and above F_{th}, the reflectance first increases, indicating that now $\omega_p > \omega$ (corresponding to $N_e > 5 \times 10^{21}\,cm^{-3}$), but then quickly decreases and increases again, apparently due to melting. After a few picoseconds the reflectance assumes a quasi-static value determined by the thickness of the metallic melt layer formed. At $4F_{th}$ explosive surface damage occurs.

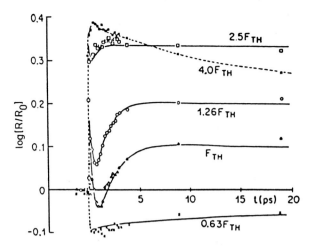

Fig.2.10. Time-resolved reflectance at $\lambda = 1 \mu m$ in crystalline Si irradiated by red 90 fs pulses of various fluences. F_{th} denotes the melt threshold fluence. [2.34]

2.2.4 Insulators

Let us now consider transparent media, assuming that $h\omega < E_g$. Even in such materials, free-carrier generation is observed at high irradiance due either to multiphoton transitions or to impact ionization.

In a *multiphoton interband transition*, n photons, the combined energy of which exceeds E_g, are absorbed simultaneously. The simultaneous absorption of n photons by an atom requires, according to the uncertainty principle, that those n photons be incident upon its cross-sectional area within a time interval $\delta t = h/\hbar\omega = 2\pi/\omega$, or one period. The probability of such an event is proportional to the photon flux $I/\hbar\omega$ raised to power n [2.35]. The rate of carrier-pair generation may be written as

$$\frac{\partial}{\partial t}N_{eh} = N\Sigma_0^{(n)}(I/\hbar\omega)^n \equiv \alpha_0^{(n)}\frac{I}{\hbar\omega} \tag{2.42}$$

where $\Sigma_0^{(n)}$ denotes the cross section for n-photon lattice absorption (with dimensions $m^{2n}s^{n-1}$) and $\alpha_0^{(n)}$ the corresponding absorption coefficient. Both quantities are related to the nonlinear dielectric function of order n appearing in (2.31) (for large n the situation becomes somewhat more complicated due to field-induced shifts in the band structure). Multiphoton ionization as a mechanism contributing to free-carrier production is of practical relevance only for small n, since the cross sections become exceedingly small for larger n. For example, $\Sigma_0^{(3)}$ in NaCl ($E_g = 8.1\,eV$) is about $10^{-80}\,cm^6 s^2$, while $\Sigma_0^{(5)}$ is $5\cdot10^{-141}\,cm^{10}s^4$ [2.36]. The corresponding ab-

27

sorption coefficients $\alpha_0^{(3)}$ and $\alpha_0^{(5)}$, calculated for ruby-laser light ($\hbar\omega = 1.78\,\mathrm{eV}$) and its second harmonic, respectively, and for an irradiance of 10^{10} W/cm^2 (which is close to the measured threshold for optical breakdown in NaCl), are 0.3 cm^{-1} and $8\cdot10^{-4}$ cm^{-1}, respectively. The relevance of multiphoton ionization in the present context is limited to short wavelengths (small n) or sub-nanosecond pulses, for which avalanche breakdown, to be treated below, is impossible. In all other cases of practical interest the dominant mechanism of carrier generation in insulators is impact ionization.

Impact ionization is the inverse process of Auger recombination: Upon colliding with a lattice atom an energetic carrier is slowed down while creating an additional low-energy carrier pair (an extra electron being knocked off the atom, leaving behind a hole). In the presence of an electric field the new carriers are accelerated again until the process repeats, and a carrier avalanche develops. The mechanism was originally proposed to explain electric breakdown in insulators by an external DC field. In the optically driven process the electrons are energized by photon absorption (the hole contribution is generally negligible due to the low mobility of holes in large-gap materials). The number of free electrons is, except for losses to diffusion and recombination, doubled in each step. If some 10^8 electrons per cm^3 are initially present in a material (this many may be expected even in reasonably pure crystals at room temperature and daylight illumination), then it takes $10^{10} \simeq 2^{34}$ additional electrons per cm^3, or about 34 generations, to bring the electron density to 10^{18} cm^{-3} which makes the crystal essentially opaque. Once this happens, formation of a microplasma [2.37] and explosive material damage is likely to occur given the high irradiance required for the electron avalanche to develop in the first place. The mechanical and thermal consequences of this phenomenon, known as "optical breakdown" will be considered within a broader context in Chap.5. Let us discuss here more specifically the conditions under which it occurs.

As long as the degree of ionization is small ($N_{eh} < N$), an energetic electron may be expected to undergo an ionizing collision as soon as it has reached a threshold ionization excess energy (measured from the bottom of the conduction band) close to E_g. The rate of creation of new electrons is then proportional to the number of existing ones times their average rate of energy gain, $\partial\langle E_e\rangle/\partial t$ (time as well as ensemble average)

$$\frac{\partial}{\partial t}N_e = N_e \frac{\partial}{\partial t}\langle E_e\rangle/E_g \,. \qquad (2.43)$$

It is evident from this that the electron density increases exponentially with time if electron losses are negligible. The essential question is thus at what rate do free electrons gain net energy inside the irradiated material. The rate of energy change experienced by an average electron as a function of the electrical field of the beam can be expressed classically as

$$\frac{\partial}{\partial t}\langle E_e \rangle = \frac{e^2 E^2}{m_e^*} \frac{\tau_e}{1 + \omega^2 \tau_e^2} - \frac{\delta E}{\tau_e} .$$

(2.44)

Here δE is the average amount of energy transferred to the lattice in a collision. (Electron-electron collisions are negligible during most of the time in which the avalanche develops). Since the electrons' kinetic energies can reach several electron volts, the collision frequencies and absorption cross sections differ from those of low-energy electrons. Collision times generally decrease with electron energy, and they become very short - of the order of 10^{-15} s - for energies of a few electron volts in any material. One consequence of the very short collision times experienced by energetic electrons is obvious from (2.44): If τ_e is of the order of 10^{-15} s, then $\omega \tau_e < 1$, and the gain term should be independent of the light frequency, up to wavelengths in the red. This is in accordance with experimental evidence: The breakdown threshold fields for large-gap substances are found to be practically the same from DC up to the ruby-laser frequency. For shorter wavelengths the gain term in (2.44) decreases like ω^{-2} and the threshold should increase accordingly. Eventually multiphoton ionization will thus have a lower threshold and dominate. The crossover wavelength depends on the band gap and the collision frequency, and is expected to be in the visible region in the case of the alkali halides [2.38].

Equations (2.43, 44), while giving the correct trends, turn out to be of little use for calculating realistic breakdown thresholds for avalanche ionization. The reason is that τ_e and δE depend on energy in different ways, which means that electrons in certain energy ranges may gain net energy while in others they lose energy. A more satisfactory approach was taken by *Holway* [2.39] who used a Fokker-Planck equation to calculate the time history of the electron distribution function under the influence of the oscillating electrical field. *Holway* found that fluctuations in electron energy can produce energetic electrons even if the average electron gains no energy at all. For illustration, Fig. 2.11 shows electron distribution functions at various times in sapphire ($E_g = 6 eV$) during irradiation by an intense ruby-laser pulse. It clearly demonstrates that avalanche breakdown is due to the "lucky" rather than to the average electrons. In view of this, what has been called the experimental "irreproducibility of the damage threshold" by *Bass* and *Barrett* [2.40] is no surprise. Indeed, the concept of "lucky" electrons implies that avalanche ionization should be of a statistical nature, at least for short pulses. *Bass* and *Barrett* found that the probability of a pulse producing breakdown varies with the field strength E inside the material like exp($-b/E$), b being a material constant related to the band gap. This was

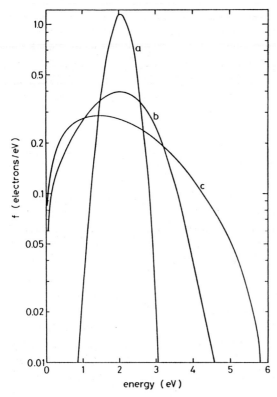

Fig.2.11. Spreading of free electron distribution in sapphire due to irradiation by a ruby laser pulse that creates an electric field of $5 \cdot 10^7$ V/cm. The steady-state distribution c, reached after about 0.1 ps, is independent of the chosen initial distribution a; b is an intermediate distribution. [2.39]

true under the condition that the radiation was tightly focused in order to reduce the probability of absorbing defects within the focal volume, and for beam powers low enough to exclude self-focusing [2.41].

It is clear that all predictions of breakdown thresholds are useless if the electrical field inside the material is not connected to the known irradiance in a simple manner. *Bloembergen* [2.42] showed that the presence of cracks, pores or inclusions with dimensions exceeding about 10 nm in a dielectric leads to local enhancements of the electrical field (and therefore to a reduction of apperant threshold irradiance), which increases with the refractive index of the material. Reduction factors of $2 \div 5$ were predicted in low-index materials such as alkali halides or optical glasses. Similarly, the apparent breakdown threshold is lower for multimode than for single-mode beams, due to transient irradiance maxima present in the former [2.43].

Finally, the threshold for optical breakdown increases substantially towards very short pulse durations, due to the finite multiplication time of the electron avalanche. Under conditions typical for optical breakdown in insulators, the avalanche build-up time is of the order of $0.1 \div 1$ ns. The breakdown threshold for shorter pulses is determined by multiphoton absorption which is an "instantaneous" process. Carrier diffusion out of the focal region also leads to an increase in the avalanche threshold for relatively long pulses of sharply focused radiation. Table A.2 lists typical values of the breakdown threshold for a choice of wavelengths, pulse durations and materials [2.44].

2.2.5 Reflection by Hot Metals

Even metals are not immune to beam-induced changes in their optical properties: their reflectance is usually observed to decrease under irradiation. The effect is thermal in nature and makes metals susceptible to thermal runaway, particularly in the infrared where reflectances tend to be high.

The bulk reflectance of metals decreases, as a rule, with temperature because the apparent electron-lattice collision time shortens. In addition, hot metals are reactive and irreversible changes in reflectance due to chemical reactions at the surface (oxidation, etc.) tend to occur, except under high vacuum conditions. Reliable experimental data on the reflectance of hot metals are scarce, but at least for the infrared we can get some information as follows. The total absorptance of a metal can be thought of as composed of three contributions, due to free electrons (fe), interband transitions (ib) and surface (surf) effects

$$(1 - R) \simeq (1-R)^{(fe)} + (1-R)^{(ib)} + (1-R)^{(surf)} \quad . \tag{2.45}$$

No general statements can be made concerning the temperature dependence of the last two terms. They depend, respectively, on details of the band structure, and on the state and the reactivity of the metal surface. However, we can relate the temperature dependence of the free-electron term to that of the DC conductivity σ_0, which is usually known. Taking the free-electron density and mass to be independent of temperature we can use (2.24, 26) to write

$$\left[\frac{1 - R(T)}{1 - R(T_0)} \right]^{(fe)} = \begin{cases} \sqrt{\sigma_0(T_0)/\sigma_0(T)} & \text{for } \omega \ll 1/\tau_e \\ \sigma_0(T_0)/\sigma_0(T) & \text{for } \omega \gg 1/\tau_e \end{cases} \tag{2.46}$$

Above room temperature the conductivities of typical metals decrease roughly linearly with temperature [2.45], and the same holds for liquid metals at not too high temperatures [2.46]. Upon melting most metals show an additional drop in conductivity. Table A.3 lists free-electron reflectances at various temperatures and compares them to experimental values at $\lambda = 1$

μm for extremely clean and for ordinary samples [2.45-47]. The relevance of the phenomenon discussed here can be judged by comparing the $(1-R)^{(fe)}$ and $(1-R)^{(exp)}$ values at 300 K. Obviously the effect is relevant only for the highly reflective metals.

In addition to the rather slight decrease in metal reflectance predicted by free-electron theory, more drastic reductions in apparent reflectance are observed in the melting and evaporation regimes. While early explanations involved effects like a metal-dielectric transition in the liquid surface layer caused by thermal expansion [2.48], or the smearing-out of the density jump at the surface by evaporation beyond the critical point [2.49], there is now little doubt that reduced reflection under laser irradiation (which may or may not indicate enhanced absorption in the condensed material) arises for one of two reasons: Mechanical deformation of the surface, possible at moderate irradiance, or plasma effects, which require irradiances sufficient for strong evaporation. Such effects are discussed in the next section.

2.3 Phase Transitions and Shape Effects

Powerful laser beams not only tend to affect the intrinsic optical properties but also the *shape* of an irradiated material. This, in turn, influences beam-solid coupling in various ways. Shape effects are almost always related to melting or evaporation.

2.3.1 Surface Corrugation

Melting a surface by laser radiation typically leaves its trace in the form of permanent ripples or corrugations. The patterns are often unrelated to the beam profile and appear even if the incident beam is smooth. A large variety of patterns have been described in the literature (a few are shown in Fig.2.12), not all of which appear to be fully understood as yet. Explaining a particular pattern requires answers to at least two questions, which may or may not be related: (i) what is the source of the pattern, and (ii) by what physical mechanism is it imprinted in the material?

In practice, by far the most frequent answer to question (i) is light scattering at material imperfections or dust particles. The scattered radiation, being coherent with the incident beam, interferes with the latter to form a pattern of modulated irradiance which now acts on the material. The basic features of many patterns can be understood from simple geometrical optics. A familiar pattern is that produced by a point-like scattering center (such as a dust particle) located in the ambient air, a distance z away from

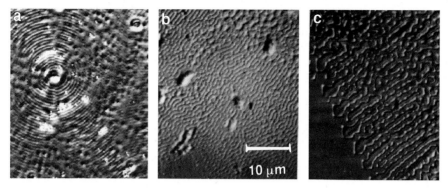

Fig.2.12. Laser-produced surface structures in (**a**) Pd, (**b**) (111)Si and (**c**) (100)Si, irridated by (**a**) 50 ns and (**b**,**c**) 200 μs Nd-laser pulses just above the melting threshold.' Pattern (**a**) is due to light scattering as described by (2.47) with $\theta = 0$, $z \simeq 0$. Patterns (**b** and **c**) are apparently unrelated to scattering

the sample surface. It consists of a system of confocal ellipses (circles for normal incidence) described by

$$r_n = \frac{n\lambda}{1 - B_1{}^2}\left[B_1 + \sqrt{1 + B_2(1 - B_1{}^2)}\right] \tag{2.47}$$

where $B_1 = \cos\phi\sin\theta$ and $B_2 = 2z/(n\lambda\cos\theta)$. Here θ is the beam incidence angle, measured from the surface normal, ϕ is the polar angle in the sample surface measured from the plane of incidence, and r_n is the distance from the vertical through the scatterer to the nth fringe. An example of such a pattern is that of Fig.2.12a. In the case of linear rather than point-like scatterers, such as scratches in the surface, fringes are parallel to the scatterer and have a spacing of $\lambda/(1 \pm \sin\theta)$ [2.50].

Often, however, patterns are more complex than simple geometrical optics would indicate. The efficiency of interference in a particlar direction depends on the state of polarisation of the incident beam. The formation of fringes with spacings of $\lambda/\cos\theta$ running parallel to the polarization direction of a p-polarized incident beam were observed on nominally smooth surfaces [2.51]. If the scattered waves propagate inside the material rather than in air, one would expect the fringe spacing to scale with λ/n_1, rather than with λ. Fringes with such spacing have, indeed, been observed in transparent dielectrics [2.52]. A generalized treatment of interference of scattered waves in terms of nonradiative "radiation remnants" propagating along the surface of a material was formulated by *Sipe* et al. [2.53].

Surface deformations are the most common and most efficient scatterers, but "latent" patterns, such as lateral variations in lattice temperature or in free-carrier density [2.54] seem able to result in some scattering even on geometrically smooth surfaces. Stimulated scattering at surface polariton waves has been observed in molten Ge [2.55] and quartz [2.56]. Whatever the origin, scattered fields from different scattering features interact, and diffraction at ripples gives rise to secondary ripples. The resulting patterns contain Fourier components at several spatial frequencies. Thus, once a fringe pattern has been physically imprinted on a surface, it can perpetuate itself coherently in subsequent, fully or partially overlapping laser shots [2.57]. This may explain the gradual build-up of periodic damage patterns that is frequently observed during repeated irradiation by pulses too weak to cause any conspicuous damage individually [2.58]. Once heavy damage has been done, surface patterns tend to deflect or scatter a large part of the incident beam, so a reduction in specular reflection is detected [2.59].

If light scattering of any kind is the answer to question (i), then question (ii) is how the modulated irradiance is physically transformed into a persisting variation of surface geometry. The basic sequence is that the material melts, undergoes deformation, and finally - after irradiation - resolidifies, making the deformation permanent. The mechanism of deformation depends on the absorbed fluence as well as on the material [2.60]. In Si or Ge irradiated near the melting threshold it appears to be related to the phase transition itself. It has been shown [2.61] that the strong decrease in absorptance upon melting of Si (Table A.1) leads to an instability at irradiances just above the melt threshold: The power absorbed by the solid is sufficient for melting, but the power absorbed by the melt is insufficient to prevent resolidifaction! Hence neither a homogeneous solid nor a homogeneous melt is a stable configuration, and a pattern of molten patches must form spontaneously. The ratio of molten to solid surface is a function of the irradiance, while the feature size is determined by the absorption length in the solid [2.62]. The melt pattern actually formed would be random for a mathematically homogeneous beam, but even the slightest modulation of irradiance, by interference or other mechanisms, will produce regularities in the pattern.

At irradiances well above the melt threshold uniform melting occurs, and lateral variations of the melt temperature become the relevant force for imprinting patterns in the material. The surface tension of liquids decreases with temperature, and liquid tends to be pulled away from hotter towards cooler regions (*Marangoni effect*). This results in meniscus-shaped deformations in the case of stationary beams and in ripple formation in the case of scanned beams [2.63]. For short pulses the modulation of surface temperature results in a modulation of both the depth and the lifetime of the melt. In layered samples the elemental distribution is found to be accordingly modulated after solidification [2.64]. Also often related to surface rippling are acoustic phenomena. Pulsed melting of Si causes an abrupt local increase in density, which acts as a strong source for acoustic waves [2.31]. Surface-acoustic or capillary waves frozen-in after irradiation have been related to surface ripples with wavelengths in the micrometer range and are unrelated to light scattering [2.65, 66]. Yet another potential source of surface ripples in the melt regime are surface oxides, which can stay solid on top of a melt layer and tend to become wrinkled like the skin on milk [2.67].

At even higher irradiance the molten surface is shaped by evaporation. Stimulated light scattering at surface corrugations driven by evaporation recoil in liquid metals was demonstrated [2.68]. On polished surfaces strong evaporation results in extensive damage, manifested optically by diffuse rather than specular reflection [2.69]. Enhanced absorption in the material may result if the depth of the surface modulation is sufficient for multiple reflections to occur.

2.3.2 Shaping Surfaces

The phenomena resulting in surface corrugation upon melting are usually seen as nuisances in laser materials processing, but they can also be put to good use. Examples include optical recording as well as various techniques that could be referred to as "laser polishing".

In *optical recording*, sub-micrometer sized holes - later to be read out as information bits - are imprinted into sensitive films by tightly focussed nanosecond diode-laser pulses. The irradiation induces a bell-shaped temperature distribution and hence a radial variation of surface tension. The molten film pulls away from the hot center to pile up at the cool rim, but the resulting depression (say, of radius r) opposes the flow by a capillary pressure $2\sigma/r$. If the irradiation persists, the two forces reach a balance and the resulting depression assumes an equilibrium shape [2.70]. However, in nanosecond recording a steady state is normally not reached and the hole shapes are influenced by viscous and inertial forces [2.71]. Gases trapped inside the film or adsorbed at the surface tend to complicate the process [2.72], but are often used intentionally to assist in the hole-formation process. The laser imprints shown in Fig.2.13 illustrate the action of radial material flow, bulging due to gas-assisted film detachment and hole opening.

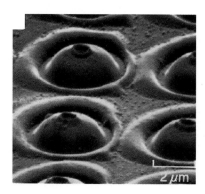

Fig.2.13. Scanning electron micrographs of laser-recorded imprints in an amorphous $In_{37}Sb_{50}Ge_{13}$ film produced with 20 μs pulses from a Ar$^+$ laser [2.72]

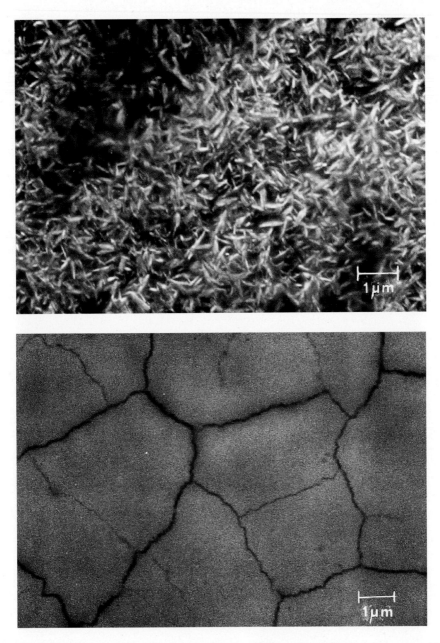

Fig.2.14. Scanning electron micrographs of a CVD-produced TiC coating on graphite before (top) and after laser remelting with a 10 ns Nd:YAG laser pulse [2.77]

Capillary forces are also a key ingredient in *laser polishing* of ragged surfaces. *Tuckerman* and *Schmitt* [2.73] demonstrated in 1985 a simple and effective method to planarize thin metal films in integrated-circuit interconnect structures. It consists of simply melting the contact by a short laser pulse, causing it to resolidify in a near-planar configuration enforced by its high surface tension. The concept has since been extended to faceted refractory films such as titanium carbide [2.74] and sapphire [2.75] and to films on soft or heat-sensitive substrates like plastics [2.76] or graphite [2.77]. The technique has a beneficial self-regulating feature as Fig.2.14 illustrates: the small-grained initial structure (TiC in the example) absorbs far more radiation than the final, smooth surface. Extended areas can be polished by subsequent overlapping pulses that will melt only the rough portions without affecting those already polished.

A rather different mechamism is at work in excimer-laser polishing of synthetic diamond films (Fig.2.15): diamond does not melt, hence the polishing effect cannot rely on surface tension. Rather, it is based on the transformation of a thin surface layer of diamond into graphite which subsequently evaporates [2.78]. Evaporation of ragged surfaces tends to smooth them because the local vapor pressure is lower in the gaps between asperi-

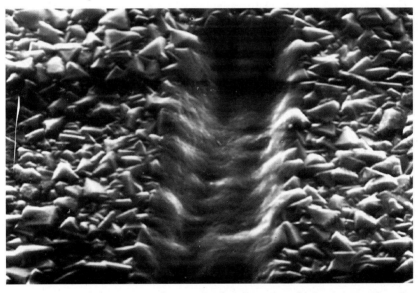

Fig.2.15. Scanning electron micrograph of a CVD-deposited polycrystalline diamond film with a polished channel produced by linearly scanned 193 nm excimer laser pulses

ties than at their tips [2.79]. Furthermore, tips tend to absorb more radiation and conduct less heat, hence they become hotter and evaporate faster. Best polishing is obtained at oblique beam incidence [2.80]. Facets with small angles to the beam normal absorb and disappear preferentially. Again, the process self-terminates when the absorption of the smooth surface falls below threshold.

2.3.3 Hole Drilling

Hole drilling, to be discussed in more detail in Sect.5.2.3, requires irradiances capable of substantial surface evaporation. The speed of material removal increases with the local irradiance, and Gaussian beams tend to form round holes with diameters roughly equal to the beam (or focal spot) diameter near the surface. However, as the hole grows deeper upon continued irradiation, it usually narrows below the surface, rather than widening as one might expect from the beam's diffraction angle. The reason for this phenomenon is "channeling" of the beam by the steep walls of the hole. As an example, Fig.2.16 shows hole profiles in Perspex drilled with CO_2-laser pulses of various durations, focused with an f/5 lens onto the surface. Channeling keeps the beam power concentrated near the axis of the hole and enables its tip to propagate, until either the irradiation stops or absorption and scattering losses in the evolving vapor jet make the process unstable (note the frayed ends of longer holes in Fig.2.16). Irregular or even curved holes tend to result in inhomogenuous materials or for unstable beams [2.81]. Optical breakdown in the vapor is promoted by the presence of molten or even solid ejected particles, particularly in the case of short, intense pulses [2.82].

In metals, beam channeling by holes yields the additional effect of reducing or even eliminating reflection losses. The hole acts as a beam trap (also known as a *keyhole*) in which the light energy is efficiently absorbed in multiple-reflection events. Effective use of "keyholing" is made in pene-

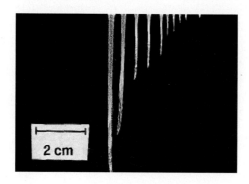

2 cm

Fig.2.16. Holes in Perspex, drilled by CO_2-laser pulses of durations 10 ms to 5 s (from left to right), illustrating self-channeling of the laser beam

tration welding by high-power CW CO_2-laser beams, to be discussed in Sect. 5.2.3. Here the beam energy is deposited deep within the molten metal by means of a cavity kept open by the evaporation pressure, enabling effective thermal penetration depths of 1 cm or more for a higher-power beam. The necessary beam powers are, however, in the kilowatt range.

2.3.4 Evaporation and Plasma Effects

The vapors formed by intense irradiation can play important roles in beam-solid coupling, particularly for infrared beams. The range of irradiances where evaporation is achieved stretches from some 10^3 W/cm^2 up to the highest irradiances realized to date (10^{15} W/cm^2 or more). It is clear that many physically distinct regimes are found in this enormous range. The following is a brief summary of phenomena related to beam-solid coupling.

At relatively moderate irradiance (below $\simeq 10^6$ W/cm^2) the vapor is tenuous and essentially transparent, but with increasing irradiance it tends to become supersaturated as it evolves from the surface. Condensing droplets of submicrometer size then lead to absorption and scattering. Apart from this, the vapor cloud is a medium of refractive index different from its surroundings and distorts the incident wavefront.

Between roughly 10^7 and 10^{10} W/cm^2 - depending on wavelength - the vapor becomes partially ionized and absorbs a substantial fraction of the laser energy. On the other hand, blackbody radiation emitted by the vapor plasma tends to be absorbed by the solid more efficiently than the laser radiation, particularly for infrared lasers. If the plasma stays close to the surface, it may actually enhance the fraction of beam energy absorbed in the solid. At irradiances somewhat higher than those producing ionization of the hot vapor, ionization may even occur in the cold ambient gas, due to optical breakdown. The breakdown plasma typically propagates as a supersonic absorption wave against the incident beam and shields the material completely. This effect seriously limits the energy deliverable by intense infrared laser beams to targets at atmospheric pressure.

At even higher irradiance (above $10^9 \div 10^{10}$ W/cm^2) the plasma, owing to its high temperature, becomes transparent and light is again transmitted to the dense surface. The ablation pressure drives a shock wave into the material which may alter its optical properties. For example, compression of semiconductors affects the band structure, in particular the band gap [2.83]. Fourfold coordinated nonmetals tend to transform into more densly packed metallic phases at high pressure. Strong shocks ionize every nonmetal. In metals a significant drop of reflectance under picosecond irradiation was predicted on the basis of a decrease in τ_e caused by loss of degeneracy in the hot electron gas [2.84]. Finally, at the highest irradiances any sharp boundary between the condensed material and the plasma disappears. Light

Table 2.1. Summary of self-induced optical coupling phenomena. (i: insulator, s: semiconductor, m: metal, ↑: increase, ↓: decrease)

Effect	Material	Spectral range	Results in
Self-focussing	i, s	$\hbar\omega < E_g$	I ↑
Thermal free-carrier generation	s	$\hbar\omega \ll E_g$	α ↑
Optical free-carrier generation	s	$\hbar\omega \gtrsim E_g$	α ↑
Avalanche breakdown	i	$\hbar\omega < E_g$	α ↑
Free-electron reflectance	m	infrared	R ↓
Heating, melting	m	any	R ↓
Melting	Si, Ge	any	R ↑, α ↑
Crystallization	a-Si, a-Ge	$\hbar\omega \sim E_g$	R ↓, α ↑
Surface corrugation	s, m	any	(R ↓)
Hole drilling	any	any	beam trapping
Vapor plasma	any	any	enhanced coupling or shielding

is absorbed at that surface where the electron density makes the plasma frequency equal to the laser frequency. Additional absorption and reduced reflectance arises in the plasma from turbulent collective motion of the electrons.

The physical conditions that produce this diversity of coupling phenomena will be discussed in some detail in Chap. 5.

To conclude this chapter, Table 2.1 presents a summary of self-induced beam-solid coupling phenomena and indicates their main effects on the optical coupling parameters. It is this set of parameters that – together with (2.10) – defines the "secondary" source mentioned at the outset. Unfortunately, as we have discussed in this chapter, the dependencies indicated are not always describable accurately by simple equations. Their practical consequences will be considered in connection with the specific processing steps discussed in the following chapters.

3. Heating by Laser Light

Heat treatments serve a wide range of purposes in today's material technology. Examples include softening or hardening of metals, "annealing" of crystals, dopant diffusion in semiconductors, compound formation in mixtures or thin-film couples, oxide-layer growth, polymerization of plastics and many others. The use of lasers as heat sources in place of furnaces is being adopted or explored in a growing number of heat-treatment processes. What advantages do lasers have to recommend them for the job?

The main advantage, obviously, is that heating can be done in a localized mode, both in space and in time. By matching the wavelength and the beam power to the optical and thermal material properties, the amount of heating can be chosen accurately to suit the needs of a process. If surface heating is required and/or volume heating is to be avoided, a large absorption coefficient and a short interaction time can be selected. Sharply delimited areas can be heated to high temperature while the remainder of the workpiece stays virtually cold. Rapid cooling of the heated material can be achieved by using short pulses or rapidly moving beams. Apart from localization, laser beams are chemically "pure" and free of inertia, can be moved easily and passed through windows to reach remote or inaccesible parts of a workpiece.

This chapter presents fundamental facts and current examples of laser solid-state thermal processing. Beam powers and pulse energies are generally moderate in this regime. The only effect of the irradiation is to raise the temperature of the material, which then reacts according to its own laws. Before discussing specific processes, we devote a section to the calculation of temperature distributions in the solid (excluding latent heat effects) for a variety of irradiation conditions and material responses.

3.1 Temperature Distributions

3.1.1 Thermalization and Heat Transport

The primary product of absorbed laser light is, strictly speaking, not heat but particle excess energy – excitation energy of bound electrons, kinetic energy of free electrons, perhaps excess phonons. The partition of the absorbed energy among the degrees of freedom of the material is not thermal at first. The degradation of the ordered and localized primary excitation energy into uniform heat involves three steps. The first step is spatial and temporal randomization of the motion of excited particles, proceeding with the collision time (i.e., momentum relaxation time) of the particles in question. This time is shorter than even the shortest laser pulses, perhaps shorter than an optical cycle. The next step, energy equipartition, tends to involve a large number of elementary collisions and intermediate states, notably in nonmetals. Several energy transfer mechanisms may be involved, each with its own characteristic time constant. For example, hot carriers in nonmetals lose energy first by phonon emission in the conduction band (Fig. 2.8), and subsequently – with a different time constant – upon recombination. Nonthermal phonon populations may be created in the process, which decay with yet another time constant. Energy equipartition is particularly slow in dielectrics and organic polymers. Intense optical excitation of such materials by UV lasers has been found to result in photodesorption of molecules without a significant temperature rise [3.1].

To describe thermal effects one usually ignores the intricacies of elementary relaxation channels and characterizes equipartition by an overall energy relaxation time τ_E, as was done in (2.41). Typical orders of magnitude for τ_E are 10^{-13} s in metals and between 10^{-12} and perhaps 10^{-6} s in nonmetals, depending on the material and on the irradiance. In semiconductors energy can be transported by hot carriers before it is given to the lattice (Sect. 2.2.3). This will be relevant whenever the carrier lifetime is not negligible compared to the laser-pulse duration.

The last step is heat flow. Once the laser energy is converted to heat it still tends to be highly localized on a macroscopic scale. The mathematical theory of heat conduction is based on the assumption that the heat flux across a plane in a solid is proportional to the local temperature gradient

$$\Phi(z_0) = -K \left[\frac{dT}{dz} \right]_{z_0} \tag{3.1}$$

where K is the thermal conductivity of the material. Accepting this for the moment, we can express the energy balance of a slab of material bounded by planes at z and $z + \Delta z$ in terms of its volumetric heat capacity c_p / V

$$\Delta t [\Phi(z) - \Phi(z + \Delta z)] = \Delta T \frac{c_p}{V} \Delta z . \tag{3.2}$$

Here ΔT is the change in temperature brought about by a net heat flux across the boundaries. Letting $\Delta z \to 0$, the bracketted term on the Left-Hand Side (LHS) can be expressed as $(\partial \Phi / \partial z) \Delta z$. Replacing the finite differences by differentials and using (3.1) then yields

$$\frac{\partial}{\partial z}\left[K\frac{\partial T}{\partial z}\right] = \frac{c_p}{V}\frac{\partial T}{\partial t} \tag{3.3}$$

which is the usual form of the heat-flow equation in one dimension. If heat is produced in the material, the power density of the heat source is added to the LHS. Before discussing solutions let us briefly examine the physical validity of (3.3) for the case of laser heating.

At issue is the validity of making Δz arbitrarily small, since this amounts to neglecting the finite mean-free path ℓ of the heat carriers (free electrons or phonons). *Harrington* [3.2] has demonstrated that in a metal most of the heat flux across a given plane is carried by electrons that had their last collision with the lattice within several mean-free lengths from the plane in question. This means that the heat flux follows the local temperature gradient only if the latter is constant over a distance of several mean-free lengths, for otherwise a significant fraction of the heat is carried by particles that "remember" a different temperature gradient. The criterion that Δz should not be made smaller than, say, $10\,\ell$ affects the validity of replacing the flux difference in (3.2) by a differential. By developing the flux difference term into a Taylor series around z one finds that (3.3) is only valid provided

$$5\ell\left|\partial^3\frac{T}{\partial z^3}\right| \ll \left|\partial^2\frac{T}{\partial z^2}\right|. \tag{3.4}$$

If not, higher-order spatial derivatives of the temperature profile must be retained to calculate the temperature distribution correctly. In metals typical absorption lengths are of the order of 10 nm, and the same holds for the mean-free paths of electrons at room temperature. The requirement of a thermal gradient constant over several mean-free paths may therefore be violated for short times. Eq.(3.3) then overestimates the heat flux away from the surface and underestimates the surface temperature. The deviation is, however, significant only for a very shallow surface region (10 nm or so) and for short times. "Linear" heat conduction theory is thus good enough for most laser applications in materials processing. Possible exceptions are laser-induced thermal surface phenomena like thermionic emission or desorption. Evidence of "nonthermal" electron distributions in tungsten irradiated by ruby-laser pulses as long as 60 ns has been reported [3.3]. "Nonlinear" heat conduction effects are also significant in plasma heating by ultrashort laser pulses (Sect. 5.4) [3.4].

3.1.2 Solution of the Heat-Flow Equation

Let us now look for solutions of the heat-flow equation with allowance for heat production by absorption of laser light. It turns out that a number of rather severe simplifications must be made to arrive at analytical solutions (more accurate numerical calculations will be discussed in Sect. 4.1.2).

First, we assume the power density of heat production to equal that of the absorbed laser light, as given by (2.10). Further, we take all material parameters to be constants[1] (a requirement to be relaxed somewhat later on). We consider a homogeneous material in the form of a slab between the planes $z = 0$ and $z = L$ (this includes the semi-infinite solid for which $L = \infty$). The material is taken to be at zero initial temperature - a uniform initial temperature is simply added to the calculated temperature. The slab is assumed to be thermally insulated, i.e., no heat flow across the boundaries is allowed. This means that the temperature distribution $T(x, y, z, t)$ must satisfy the condition

$$\frac{\partial T}{\partial z} = 0 \quad \text{for} \quad z = 0 \quad \text{and} \quad z = L \quad \text{at all times.} \tag{3.5}$$

The laser beam is incident onto the plane $z = 0$ and taken to be of cylindrical symmetry. The heat-flow equation can now be written as

$$\frac{\partial T}{\partial t} = \kappa \nabla^2 T + \frac{\alpha I_a(x, y, t) V}{c_p} e^{-\alpha z} \tag{3.6}$$

where $\kappa = KV/c_p$ is the heat diffusivity, and I_a is the unreflected part of the incident irradiance.

A mathematical method well adapted to the laser-heating problem is the Green's function technique [3.5]. The Green's function $\mathbf{g}$ gives the temperature distribution produced by an "instantaneous" heat source of unit energy and has dimensions [K/J]. Once g is known, the transient temperature distribution for a continuous heat source of power

$$P_a(t') = \iint I_a(x', y', t') dx' dy'$$

is found from

$$T(x, y, z) = \int_0^t P_a(t') g(x, y, z, x', y', z', t-t') dt' \tag{3.7}$$

---

[1] Suitably averaged specific heats can be estimated from the data of Table A.4.

where x,y,z,t and x',y',z',t' are the coordinates of the field and source points, respectively. We shall treat two kinds of lateral irradiance distributions here, the uniform and the Gaussian distributions. The former describes one-dimensional heating by a large-area beam where lateral heat flow can be neglected while the latter applies to a focused beam of fundamental-mode radiation. Further, we consider the limiting case of a vanishing absorption length (referred to as a surface source), as well as a finite absorption length (penetrating source). Green's functions for these source geometries are given in the Appendix B. In the following we present a number of solutions for the slab and for the semi-infinte solid, for use in later discussions. We shall only consider sources of constant power, being switched on at $t = 0$. For notational convenience, we introduce the diffusion length $\delta = 2(\kappa t)^{1/2}$.

For a *uniform surface source* of irradiance $I_a = P_a/S$ we have

$$T(z,t) = \frac{I_a \delta}{K} \sum_{n=-\infty}^{\infty} \mathrm{ierfc}\left(\frac{2nL - z}{\delta}\right). \tag{3.8}$$

In particular, for the semi-infinite solid

$$T(0,t) = I_a \frac{\delta}{K\sqrt{\pi}}. \tag{3.9}$$

The general expression for the *uniform penetrating source* is

$$T(z,t) = \frac{I_a}{4K} \int_0^\delta \exp[-(\alpha\beta/2)^2]\Upsilon_2\, d\beta. \tag{3.10}$$

Closed solutions are found for the semi-infinite solid

$$T(z,t) = \frac{I_a}{K}\left\{\delta \cdot \mathrm{ierfc}(z/\delta) - (1/\alpha)e^{-\alpha z} + (1/2\alpha)e^{(\alpha\delta/2)^2}\right.$$
$$\left. \times\left[e^{-\alpha z}\mathrm{erfc}(\alpha\delta/2-z/\delta) + e^{\alpha z}\mathrm{erfc}(\alpha\delta/2+z/\delta)\right]\right\} \tag{3.11}$$

and

$$T(0,t) = \frac{I_a}{K}\left\{\frac{\delta}{\sqrt{\pi}} - \frac{1}{\alpha}\left[1 - e^{(\alpha\delta/2)^2}\mathrm{erfc}(\alpha\delta/2)\right]\right\}. \tag{3.12}$$

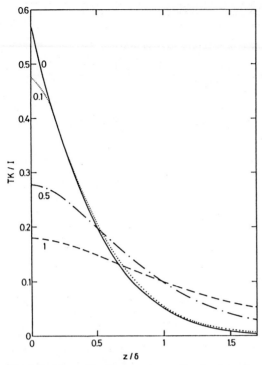

Fig. 3.1. Normalized temperature distributions for uniform heat sources as a function of normalized depth. Numbers indicate the ratio of the absorption length α^{-1} to the diffusion length $\delta = 2\sqrt{\kappa t}$

Temperature profiles for uniform sources with various values of $\alpha\delta$, calculated from (3.8) (with $L = \infty$) and from (3.11), are shown in Fig. 3.1. Note that going from zero absorption length to an absorption length of one tenth of the diffusion length lowers the surface temperature markedly, but barely affects the profile deeper within the solid. The temperature at depths beyond one absorption length is thus well described by the surface source provided that $\delta \gg 1/\alpha$ (in the case of metals this holds for times exceeding about 0.1 ns, i.e., whenever the present linear theory holds). In the other extreme, for $\delta \ll 1/\alpha$, the temperature distribution essentially follows the exponential absorption profile.

The general expression for the *Gaussian surface source* is

$$T(r, z, t) = \frac{P_a}{2K\pi^{3/2}} \int_0^\delta \exp\left(\frac{-r^2}{\beta^2 + w^2}\right) \Upsilon_1 \frac{d\beta}{\beta^2 + w^2} . \tag{3.13}$$

A closed solution is found for the slab at steady state $(t = \infty)$

$$T(0, z, \infty) = \frac{P_a}{4Kw\sqrt{\pi}} \sum_{n=-\infty}^{\infty} \exp\left[\frac{(2nL - z)^2}{w^2}\right] \mathrm{erfc}\left[\frac{2nL - z}{w}\right] \quad (3.14)$$

Simple expressions for the semi-infinite solid are

$$T(0, 0, t) = \frac{P_a}{Kw\pi^{3/2}} \arctan(\delta/w) , \quad (3.15)$$

$$T(0, 0, \infty) = \frac{P_a}{2Kw\sqrt{\pi}} , \quad (3.16)$$

$$T(r, 0, \infty) = \frac{P_a}{2Kw\sqrt{\pi}} \exp(-r^2/2w^2) I_0(r^2/2w^2) . \quad (3.17)$$

In (3.17), which was first derived by *Lax* [3.6], I_0 denotes the modified Bessel function of order zero. Note that the Gaussian source, as any kind of laterally confined source, leads to a finite steady-state temperature distribution, in which heat flow just balances the input from the source (the indefinitely increasing temperature (3.8) is due to a source of infinite power). Comparing (3.15) with (3.16) shows that 90% of the steady-state temperature is reached after a time $t \simeq 10w^2/\kappa$. Finally, the general expression for the *Gaussian penetrating source* is

$$T(r, z, t) = \frac{\alpha P_a}{4\pi K} \int_0^\delta \exp\left[\frac{-r^2}{\beta^2 + w^2} + \frac{\alpha^2 \beta^2}{4}\right] \Upsilon_2 \frac{\beta \, d\beta}{\beta^2 + w^2} \quad (3.18)$$

for which only solutions in terms of tabulated functions are known [3.6].

3.1.3 Cooling

To include cooling of the material after the laser pulse, the integration in (3.7) is extended beyond the pulse duration. Consider the uniform surface source with a rectangular temporal pulse shape of duration t_p and an irradiance $I_a = P_a/S$. The temperature distribution during and after the pulse is described by

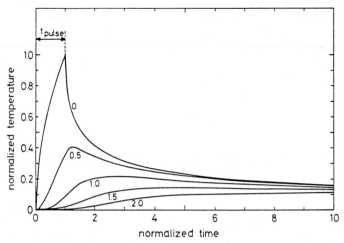

Fig. 3.2. Normalized temperature as a function of normalized time t/t_p at various depths $z/2(\kappa t_p)^{-1/2}$ for a pulse of rectangular temporal shape

$$T(z,t) = P_a \int_0^t g_{us} dt' = (2I_a/K)(\kappa t)^{1/2} i erfc(z/2\sqrt{\kappa t})$$

$$- [t > t_p]\sqrt{\kappa(t - t_p)} \, i \, erfc[z/2\sqrt{\kappa(t - t_p)}] \qquad (3.19)$$

where the symbol $[t > t_p]$ equals one if $t > t_p$ and zero otherwise. Normalized temperature-vs-time curves at various depths computed from (3.19) are shown in Fig. 3.2. Cooling curves for other source geometries and pulse envelopes are obtained analogously.

3.1.4 Moving Sources

There are two types of situation in which the medium is moving with respect to the heat source: Either the laser beam is scanned across the sample surface, or the light absorbing zone, with a stationary beam, moves into the target as a result of material removal. Formally, the motion of the medium with respect to the source is allowed for by replacing the coordinates (x, y, z, t) in (3.7) by "delayed" coordinates. For example, for a motion at the velocity u parallel to the x axis, the field coordinates are $([x-u(t-t')]$, $y, z, t)$. With this substitution any desired temperature distribution caused by a moving beam can be derived by the procedures used above. We only consider surface sources of constant power and velocity, heating a semi-infinite solid ($L = \infty$).

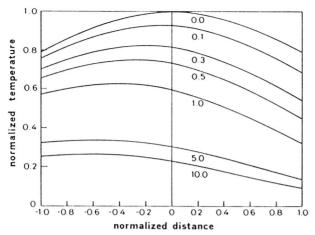

Fig. 3.3. Normalized surface temperature versus normalized distance from the beam axis, $(x-ut)/w$, for a continuous beam scanned at constant velocity u along the x axis. Numbers indicate values of the normalized velocity $uw/2\kappa$ [3.8]

For the *Gaussian source scanned along the x axis* we get

$$T(x, y, z, t') = \frac{P_a}{\pi^{3/2} K} \int_0^\infty \exp\left[-\frac{(x-ut+u\beta^2/4\kappa)^2 + y^2}{\beta^2 + w^2} - \frac{z^2}{\beta^2}\right] \frac{d\beta}{\beta^2 + w^2} .$$

(3.20)

Equivalent expressions have been given in the literature [3.6, 7]; an elliptical beam was treated by *Nissim* et al. [3.8]. For illustration, Fig. 3.3 shows surface temperature profiles along the x axis in a moving frame for a circular beam scanning at various velocities, and normalized to the steady-state temperature (3.16) at the center of a stationary beam. The scaling of the surface temperature with P_a/w shown by (3.16) turns out to be preserved at a nonzero scanning velocity and is a characteristic feature of heating by a laterally scanned beam.

The steady-state distribution for a *uniform surface source* moving *in the z direction* at velocity u is obtained form (3.7) as

$$T(z, t) = I_a \frac{\kappa}{Ku} e^{-\zeta u/\kappa}$$

(3.21)

where $\zeta = z - ut$ is a coordinate parallel to z and originating at the source. It is assumed here that the physical surface is always at $z = \zeta$, due to evaporation of the material in the region between 0 and ζ, as we shall discuss in detail in Chap. 5.

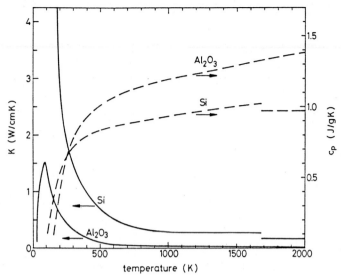

Fig.3.4. Thermal conductivity K and specific heat c_p as a function of temperature for single-crystalline intrinsic Si and polycrystalline Al_2O_3

3.1.5 Variable Parameters

The analytical approach to heat flow, elegant and intuitive as its results may be, rests on the rather unrealistic assumption of temperature-independent material properties. The expressions derived above are adequate to reproduce the essential features of a temperature distribution, but more than qualitative agreement with experiments cannot usually be expected. Both K and c_p, and hence κ, are often quite strong functions of temperature, particularly in nonmetals. Figures 3.4,5 show K and c_p as a function of temperature for sapphire and Si, and for Al and Cu, respectively. We shall discuss numerical means to obtain accurate solutions of the heat-flow equation with arbitrarily variable parameters in Sect.4.1, but let us, for the moment, consider methods to improve on the realism of the analytical approach without sacrificing all its virtues. A classical method consists in allowing the conductivity to depend on temperature while keeping V and κ constant. Note that the constancy of κ implies proportionality between c_p and K. As first proposed by G. Kirchhoff, we introduce a new variable

$$\theta = \frac{1}{K_0} \int_{T_0}^{T} K(T')dT' \tag{3.22}$$

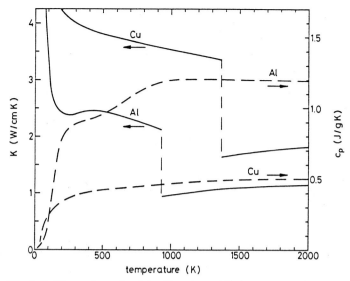

Fig.3.5. Thermal conductivity K and specific heat c_p as a function of temperature for Cu and Al

where K_0 is the conductivity at some reference temperature T_0. The heat flow equation in terms of θ reads

$$\frac{\partial \theta}{\partial t} = \kappa \nabla^2 \theta + J(x, y, z, t)\kappa/K_0 \; . \tag{3.23}$$

If J does not itself depend on temperature, then the form of (3.6) is preserved, and all solutions given above can be taken over by replacing T by θ and K by K_0. The transformation from θ back to T requires inversion of (3.22), which can be done on a desk calculator. An alternative procedure is to fit K(T) to an analytic expression for which the integral (3.22) yields a closed-form solution. For example, the conductivity of Si is well represented by the form K(T) = a/(T−b) [3.8] (for Si the use of (3.22) is not particularly suitable given that κ varies with temperature more strongly than K in the temperature range of interest).

We note in passing that the relation (3.22) may also serve to transform the temperature profiles given above into a normalized, material-independent form. For instance, with reference to (3.19), if we denote the surface temperature at the end of the pulse by

$$\theta' = \theta(0, t_p) = \frac{2I_a}{K_0} \sqrt{\kappa t_p / \pi} \tag{3.24}$$

then we can define the dimensionless coordinates

$$\theta^* = \theta/\theta' \; ; \;\; z^* = z/2\sqrt{\kappa t_p} \; ; \;\; t^* = t/t_p \tag{3.25}$$

in terms of which (3.19) reads

$$\theta^*(z^*,t^*) = \sqrt{\pi t^*}\, i\,\mathrm{erfc}(z/\sqrt{t^*}) - [t > t_p]\, \sqrt{\pi(t^*-1)}\, i\,\mathrm{erfc}(z^*/\sqrt{t^*-1}) \; . \tag{3.26}$$

This equation holds for any material, provided the temperature dependence of V and κ can be neglected.

The next step towards increased computational sophistication would be to solve the integral (3.7) numerically. The concept of the instantaneous heat source then offers the possibility of including further phenomena, such as self-induced coupling effects or more realistic pulse shapes, by adjusting the pertinent parameters after each step in the integration. This procedure yields a reasonably realistic description of laser-beam heating with relatively moderate computing expenditure. Still beyond the reach of this method are, however, phase transitions involving latent heat, such as melting and solidification.

3.1.6 Impact of Absorption Phenomena on Temperature

The coupling parameters R and α, as discussed in Chap. 2, tend to undergo dynamical changes during irradiation (Table 2.1). The question we want to address here is how these changes affect sample heating. We shall take a qualitative approach based on analytical formulas.

As an example, consider uniform surface heating (stationary beam) of a material the reflectivity of which decreases linearly with temperature. Setting $R(T) = R_0 - bT$, and taking all other parameters to be constant, the surface temperature is found to be [3.9]

$$T(0,t) = \frac{1-R_0}{b} \frac{1}{[1 + \mathrm{erf}(B)]\exp(B^2) - 1} \tag{3.27}$$

where

$$B = Ib\sqrt{\kappa t}/K \; .$$

This expression increases faster than exponentially with time, rather than just as $\sqrt{t}$, predicted by (3.9) for a constant reflectivity. Similar behavior is found in the case of an absorption length which decreases with temperature,

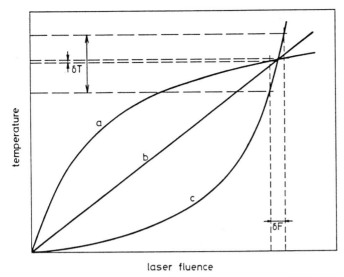

Fig.3.6. Schematic increase of material temperature with incident laser fluence. The shape of the curve (a, b, c) determines the variation in final temperature, δT, caused by a variation in fluence δT

as observed in semiconductors at photon energies near the band gap [3.10]. Such accelerated heating illustrates an important property of many of the effects listed in Table 2.1, namely the tendency to amplify by positive feedback. The problem also exists, if somewhat less pronounced, for scanned CW beams [3.11].

In most applications of laser material processing the sample temperature is a critical parameter which must be kept within narrow limits in order to ensure the desired result. A behavior like that of (3.27) is unacceptable in view of the lateral beam profiles as well as the reproducibility of real lasers. The crucial question is therefore, what temperature variations are induced by a given amount of variation in beam energy or irradiance?

The answer depends on the relation between the temperature and the pulse fluence. Figure 3.6 exhibits three schematic temperature-vs-fluence curves. For curve (b) the temperature rises linearly with the fluence, for curve (a) the rate of temperature rise decreases with fluence, and for curve (c) the reverse is true. The final temperature is the same in all cases, but obviously a given variation in beam fluence (either due to the lateral beam profile or to energy fluctuations from pulse to pulse) causes a large variation in temperature for case (c), whereas for case (a) it does not. From a practical point of view, clearly case (a) is the desirable one. Which of the cases in Fig.3.6 represents a given situation depends on absorption phenomena as well as on heat flow.

Consider a pulse of fluence F, incident on the surface of a sample of thickness L, creating heat over the thickness $1/\alpha$. During the pulse of duration t_p (thought to be fixed), heat diffuses over a dimension $\delta_p = 2(\kappa t_p)^{1/2}$; let us assume that $\delta_p < L$ (otherwise the following discussion applies with δ_p replaced by L). Two limiting cases may be distinguished:

(i) **If $\alpha^{-1} \ll \delta_p$**, then the surface temperature at the end of the pulse is, from (3.9)

$$T(0, t_p) \simeq 1.13 F(1-R) \frac{V}{c_p} \sqrt{\kappa t_p} \,. \tag{3.28}$$

In this case the slope T(F) depends on the behaviors of both R and c_p. Since c_p increases with temperature, T(F) tends to saturate, corresponding to case (a) above, unless the reflectivity decreases during the interaction, as is often the case in metals.

(ii) **If $\alpha^{-1} \gg \delta_p$**, then heat is created over a large depth. Expanding (3.12) in terms of $\alpha\delta$ and neglecting higher-order terms yields

$$T(0, t_p) \simeq \alpha F(1-R) \frac{V}{c_p} \,. \tag{3.29}$$

In this case clearly the behavior of α as a function of temperature or fluence is crucial. A small but increasing absorption coefficient results in a very unfavorable type (c) heating curve, at least as long as the condition $\alpha^{-1} \gg \delta_p$ holds. This situation occurs in nonmetals at photon energies near or below the band gap. Controlled heating is quite impossible in such cases.

Measures by which type (c) heating can be avoided in practical situations follow from the above considerations. Examples are an appropriate choice of laser wavelength (perhaps by frequency doubling), the use of absorbing layers or antireflective coatings on highly reflective metals, or the provision of an adequate initial absorption coefficient, by preheating, carrier injection or auxiliary short-wavelength illumination, in nonmetals. In addition, the beam profile can be smoothed by suitable devices to reduce the variance δF. For example, *Cullis* et al. [3.12] described a curved quartz rod beam guide through which they passed their ruby-laser pulse to obtain an essentially square beam profile. Similar results were obtained in the author's lab with a simple aluminum beam guide with a conical bore, the diameter and angle of which are matched to the beam size. The device, which can be fitted with a ground glass plate at its entrance, also brings about a slight focusing effect. The smoothing action of such devices relies on spatial and temporal randomization of the pulse energy by means of many different optical path lengths. The smoothed profiles still contain transient irradiance maxima, but these tend to be short-lived and localized enough to be averaged out by heat flow.

3.2 Heat-Treatment Processes

3.2.1 Annealing with Laser Beams

The typical purpose of annealing is the promotion of a thermally activated reaction like crystallization or bulk diffusion. In such reactions the reaction frequency per atom or molecule follows the Arrhenius law

$$\nu(T) = \nu_0 \, e^{-Q/kT} \tag{3.30}$$

where T is the absolute temperature, and Q is the activation energy per particle (e.g., the energy needed to break one kind of bond in order to allow the formation of another, or the strain energy of a lattice deformation that lets a diffusing atom "squeeze" itself across a lattice plane). The quantity ν_0 can be thought of as an "attempt frequency", of the approximate order of the thermal vibration frequency kT/h, while the exponential factor accounts for the probability that an attempt is successful. Quite often Q is only approximately constant and different values must be used in different temperature regimes. Expression (3.30) is valid for a reaction taking place under homogeneous and isothermal conditions. If the reaction goes on for a time t, the total amount of reaction is proportional to

$$n = \nu t = \nu_0 t \, e^{-Q/kT} \; . \tag{3.31}$$

Now, conditions in laser processing are usually far from being uniform and isothermal, and the simple law (3.31) thus appears inadequate. Fortunately, it can be adapted to nonisothermal processing by simple reasoning as follows: In solid-state processing the variation of temperature with time is typically rather slow compared to the time constants of elementary processes. It is thus reasonable to replace (3.31) by

$$n(t) = \nu_0 \int_{-\infty}^{t} e^{-Q/kT(t')} \, dt' \tag{3.32}$$

where T(t) describes the local temperature history. It is now a straightforward matter to determine the amount of reaction due to an irradiation event by inserting the appropriate temperature function and solving the integral. Yet the procedure can be further simplified. Consider a reaction taking place in a strongly absorbing thin film irradiated by a square pulse of duration t_p. If the film thickness is small compared to the thermal diffusion length then its temperature can be taken to be uniform and equal to the surface temperature. Using (3.19) and allowing for a finite ambient temperature T_0 we may write

$$T(t) = T_0 + \Delta T_{max}\left[\sqrt{t/t_p} - [t > t_p]\sqrt{t/t_p - 1}\right] \qquad (3.33)$$

where $\Delta T_{max} = (2I_a/K)(\kappa t_p/\pi)^{1/2}$ is the maximum temperature rise, reached at $t = t_p$. The expression resulting if (3.33) is inserted into (3.32) can be shown to be well approximated by the form of (3.31) [3.13], i.e.,

$$n(t) \simeq \nu_0 t_{eff} e^{-Q/kT_{max}} \qquad (3.34)$$

where $T_{max} \equiv T_0 + \Delta T_{max}$, and where t_{eff} is an "effective" reaction time that depends on the temperature. It is given by

$$t_{eff} = t_p(2k\Delta T_{max}/Q)(T_{max}/\Delta T_{max})^2 . \qquad (3.35)$$

The approximations made in deriving (3.34) are based on the strong temperature dependence of (3.30), which allows one to neglect reactions occurring far from the maximum temperature. Similar reasoning can be applied in the case of a laterally scanned beam [3.14][2] if the dwell time $2w/u$ exceeds the time $10w^2/\kappa$ required for the surface temperature to reach the steady-state value (3.17). The temperature at a point on the surface through which the beam moves at $t = 0$ is given by (3.17) with the coordinate r replaced by ut. Inserting this into (3.32) again yields a result that can be represented by (3.34), but with ΔT_{max} now given by (3.17), and t_{eff} by

$$t_{eff} = \frac{2w}{u}\sqrt{\pi k\Delta T_{max}/2Q}\,\frac{T_{max}}{\Delta T_{max}} . \qquad (3.36)$$

This result can be generalized to the case in which an extended area is treated by a series of partially overlapping parallel scans [3.15]. Integrating the amount of reaction by subsequent scans again yields a result of the form (3.34), with ΔT_{max} given by (3.17) and t_{eff} by

$$t_{eff} = t\,\frac{\pi w^2}{S_{sc}}\,\frac{2k\Delta T_{max}}{Q}\left[\frac{T_{max}}{\Delta T_{max}}\right]^2 \qquad (3.37)$$

t being the total scan time and S_{sc} the scanned area. Note that (3.37) is equivalent to (3.35), i.e., in the approximation used here, scanning a given area with a focused beam is equivalent to irradiating the same area with a single pulse of appropriate duration.

---

[2] *Gold* and *Gibbons* considered the special case of Si with a temperature-dependent conductivity of the form $K(T) = a/(T-b)$

Although we have neglected temperature gradients inside the material, it is clear that the irradiated surface will be somewhat hotter than the regions beneath, and reactions will usually start here and proceed inwards. The main quantity of interest is the thickness of material reacted in an irradiation event. To calculate this we need, in addition to the above formulas, a relation between the reaction rate v and the velocity of the interface separating reacted and unreacted material. This relation is characteristic of the type of reaction. In *interface-limited reactions* the interface moves at a velocity given by av, a being a molecular diameter (Sect. 4.1.4). The total thickness of material transformed during an irradiation is then

$$\Delta z = a v_0 t_{eff} e^{-Q/kT_{max}} .$$ (3.38)

Examples of such reactions are epitaxial growth (crystallization) of amorphous semiconductor layers as well as certain compound-forming reactions in layered binary thin-film structures. *Diffusion-limited reactions*, in contrast, show a $\Delta z \propto \sqrt{t}$ rather than a $\Delta z \propto t$ dependence and can be described by a law of the form

$$\Delta z^2 = B t_{eff} e^{-Q/kT_{max}}$$ (3.39)

where B is a material constant. Reactions of this type are limited by the rate of transport of unreacted atoms to the interface, and it is the diffusivity that is responsible for their Arrhenius behavior. Most compound forming solid-state reactions, including oxide growth, are diffusion- rather than interface limited.

3.2.2 Thermal Stress

Material which is heated, expands. The transient and nonuniform temperature distributions resulting from laser heating cause different zones of the material to expand by different amounts and thermal stress to develop. While the stress is small it is absorbed by the elasticity of the material. If the stress exceeds a certain level, the material yields and defect formation, cracking or plastic flow occur. Thermal stress effects are probably the most serious difficulty in the way of a successful beam-annealing application.

A mathematical description of thermal stresses caused by nonuniform heating is complex even if strongly simplifying assumptions are made, and is beyond the scope of this volume. A systematic development of the theory of thermal stress with solutions for various geometries can be found in the book by *Boley* and *Weiner* [3.16]. Under some simplifying assumptions persistent deformation or damage can be assumed to occur if at some point the local stress exceeds the yield stress of the material. The yield stress is often anisotropic and, in general, it depends on the local temperature as well as on the strain rate.

In single-crystalline Si yielding occurs preferentially by slip along (111) planes. The yield stress is related to the activation energy for the formation of slip dislocations and decreases in an exponential fashion with temperature. The amount of damage created by stress beyond the yield point ranges from isolated dislocations reducing carrier lifetimes to macroscopic deformation or peeling of films, depending on the maximum temperature and on the degree of inhomogeneity of the temperature distribution. It appears that short pulses are less liable to induce damage due to the finite time required for the nucleation of slip dislocations. *Rozgonyi* and *Baumgart* [3.17] concluded that pulses shorter than 50 ms do not induce slip in Si, even if the melting temperature is reached. On the other hand, powerful pulses can lead to thermal stress damage extending far beyond the directly irradiated region where damage, e.g., by cratering, is most obvious [3.18].

To model stress development during CW laser-induced re-growth in single-crystalline Si, *Correra* and *Bentini* [3.19] calculated stress distributions produced by a scanned line-shaped heat source. They found that local stresses can change rapidly from tensile to compressive as the beam moves across the surface of the sample, and they were able to relate the topographical distribution of stress-induced defects to the irradiation condition producing them. If heating is rapid, large temperature gradients are also created accross the thickness of a wafer. *Cline* [3.20] estimated that thermal deformation in Si wafers should occur for heating rates exceeding 8000 K/s. He also investigated deposited thin films of Si and found that thermal stress due to nonuniform heating causes plastic deformation of the film for lateral temperature differences as small as 10 K. Even totally uniform heating by a substantial amount can, of course, produce thermal deformation in thin film structures if the thermal expansion coefficients of the film and the substrate are different.

In metals, the energies for formation and propagation of dislocations are much smaller than in Si. Extensive defect formation and deformation is already observed far below the melting point and even for pulses in the ns regime [3.21]. In single-crystalline specimens defects are found to be strongly orientation-dependent [3.22, 23]. In polycrystalline materials subjected to repetitive heating cycles, as, e.g., in high-power laser mirrors, intergranular slip and fatigue cracks are known to lead to gross structural damage, without the material ever being molten.

3.2.3 Crystallization of Amorphous Semiconductor Layers

In modern semiconductor technology amorphous semiconductor layers, typically some 100 nm thick, arise as a result of ion-implantantion of dopants into single-crystalline wafers. Alternatively, amorphous layers can be pro-

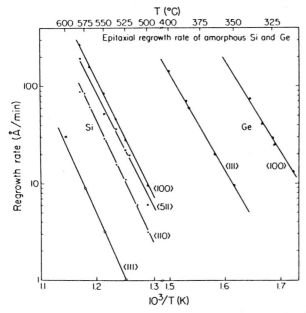

Fig.3.7. Regrowth velocity as a function of temperature in self-ion implanted amorphous Si and Ge for various crystallographic directions [3.24]

duced by vapor deposition on suitable substrates. Upon heating, the amorphous material becomes thermodynamically unstable and crystallizes spontaneously. Crystallization is unidirectional only if it is epitaxial, i.e., if the crystallized layer grows from one single seed - usually the substrate - rather than from randomly formed nuclei.

The Arrhenius law holds remarkably well in this case. Figure 3.7 shows epitaxial growth velocities in various crystallographic directions on a logarithmic scale versus the inverse temperature in ion-implanted Si and Ge wafers, as measured during isothermal furnace annealing [3.24]. Note that the prefactors ν_0 are sensititve to the orientation (they also depend on the density and nature of impurities present in the semiconductor), whereas the activation energies are not. *Olson* et al. [3.25] used a scanned laser to determine regrowth velocities in amorphous Si also at higher temperatures. Interface velocities were inferred from time-resolved reflectivity measurements, taking advantage of the different refractive indices of crystalline and amorphous Si. The experiment showed no deviation of the activation energy from its low-temperature value up to temperature close to the melting point of crystalline Si.[3]

---

[3] This contrasts to thermodynamic considerations according to which amorphous Si should melt at a temperature several hundred degrees lower than crystalline Si (Chap. 4).

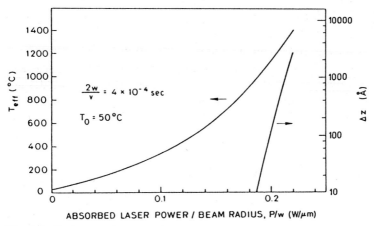

Fig.3.8. Peak temperature and regrown layer thickness in self-ion implanted Si as a function of absorbed power per unit beam radius [3.15]

To illustrate typical parameters in scanned-beam annealing, Fig. 3.8 exhibits peak temperature and regrown layer thickness as a function of absorbed laser power per unit beam radius in self-implanted amorphous Si, calculated from (3.38) using $a\nu_0 = 3.2 \cdot 10^4$ m/s and $Q = 2.35$ eV [3.15]. The main advantage of scanned-beam annealing of semiconductors is that it allows higher effective growth temperatures and shorter processing times than conventional furnace annealing. The higher temperature generally leads to better crystal quality, while the shorter annealing time virtually eliminates the unwanted redistribution of dopant atoms by solid-state diffusion [3.26]. Wafers are often preheated to a few hundred degrees before laser scanning in order to improve absorption and controllability of the process. Carefully choosing the beam power and scan speed is essential to avoid local overheating which may result in slip dislocations [3.27], or even in local melting causing surface corrugation. The concentration of electrically activated dopant atoms like As can exceed the equilibrium solid-solubility limit, but the excess dopants are deactivated if the material is subsequently furnace annealed [3.28]. Work with compound semiconductors like GaAs has been less successful, mainly because of the tendency of such materials to decompose at high temperatures, and because of thermal stress problems [3.29]. A general weakness of the technique of solid-state, as opposed to liquid-state, epitaxy (discussed in Chap. 4), is its extreme sensitivity to impurities like O, N or C, which lead to defect formation or may prevent single-crystal growth altogether even at relatively small concentrations.

Scanned-beam annealing techniques have also been used to produce single-crystalline semiconductor sheets on amorphous substrates from which no epitaxial growth is possible. The interest in this application derives from current efforts towards three-dimensional integration of VLSI circuits. *Geis* et al. [3.30] used a process they termed "graphoepitaxy", in which fine artificial relief gratings were etched into fused silica substrates to serve as templates for oriented crystal growth. They obtained large oriented grains from CVD deposited amorphous Si films on such substrates upon scanned Ar laser annealing. Related attempts have used predefined amorphous islands to control crystal alignment during laser scanning [3.31]. The island boundaries also act as dislocation sinks and allow relief of thermal misfit stress between the substrate and the film.

The maximum achievable grain size in this as well as in other annealing techniques is limited by random nucleation. The rate of random nucleation has a stronger temperature dependence than the epitaxial growth rate (Sects. 4.1.4, 5), and lower annealing temperatures should favor larger grains. Stress and interface effects seem to limit grain sizes achievable by uniform (furnace) heating to some 10 μm or less, however [3.32]. Localized heating by laser scanning has the crucial advantage of allowing suppression of nucleation by means of suitably taylored temperature profiles. For example, heating of the amorphous material can be concentrated to the immediate neighborhood of the desired growth interface. Shaping the thermal gradient influences the growth direction of unwanted crystallites. *Biegelsen* et al. [3.33] reported that by replacing the circular beam profile by a crescent-shaped one, unwanted crystallites could be forced to grow out of the beam path, enabling continuous growth of one single crystallite by the scanned beam. Reported crystallites dimensions by laser scanning can exceed 100 μm [3.34].

An interesting aspect of solid-state regrowth is the role played by the latent heat liberated upon crystallization. The latent heat contributes significantly to heating in materials like Si and Ge, and it may in fact make the transformation self-promoting, an effect known as *explosive crystallization*. It is most readily observed in films on poorly heat-conducting substrates. Figure 3.9 exhibits two different morphologies of crystallized amorphous Ge on a silica substrate, scanned with a focused Kr laser beam at slow (left frame) and fast (right frame) scanning speed. Similar morphologies can be observed in Si. The track of small-grained polycrystalline material shown on the left part of the figure results from normal solid-state crystal growth, whereas the large-grained crescent-like structures on the right result from a periodic explosive growth mode. The periodicity of the pattern for a given material is determined by the power and scan speed of the beam as well as by the ambient temperature [3.35]. Rapid explosive growth can proceed at velocities exceeding 1 m/s [3.36] and seems to involve a narrow molten film separating crystalline and amorphous regions [3.37]. Slower self-promoted crystal growth not involving a liquid layer has also been observed. Which one of the various possible growth modes actually materializes depends on the beam power, the scan speed and the substrate temperature as well as on the homogeneity of the amorphous film [3.38].

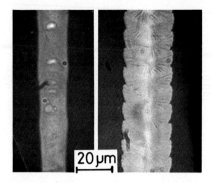

Fig.3.9. Optical transmission micrographs showing two typical morphologies in scanned laser annealed amorphous Ge on a glass substrate. *Left*: scanning speed below 2 cm/s, *right*: above 2 cm/s. Courtesy G. Badertscher, Bern.

20 μm

3.2.4 Compound Synthesis

Scanned laser beams have also proven useful in the formation of compounds from binary layered structures. The best-studied application to date is the growth of metal silicides for use as ohmic or Schottky-barrier contacts to Si devices [3.39]. A metallic film, of the order of 100 nm thick, is deposited on a Si wafer and heated by a pulsed or CW scanned beam. Upon heating the elements react spontaneously, starting from the original metal-Si interface, by forming a binary compound. The composition of the compound formed, as well as the kinetics of its growth, are characteristic of the metal. It is known from isothermal furnace annealing that in systems with several equilibrium compounds one particular compound always forms first, but only empirical rules are available to predict the phase forming first. A well-known rule, due to *Walser* and *Béné* [3.40], states that the first silicide to grow is the one with the highest congruent melting point next to the deepest eutectic. The rule has been generalized [3.41] to cover other alloy systems.

The kinetics in a number of cases follows (3.38), i.e., the compound layer thickness grows linearly with time for a constant temperature. Examples of systems behaving this way are Cr-Si or Mo-Si [3.42]. However, most compound forming solid-state reactions, including oxide growth, are diffusion rather than interface limited and grow according to (3.39). As an example of laser-induced diffusion limited compound formation, Fig.3.10 displays the absorbed laser power as a function of the substrate temperature T_0, required to grow 10 and 100 nm thick layers of the silicide Pd_2Si from a Pd layer on top of a Si substrate, subjected to partially overlapping scans (calculated from (3.39) with $B = 3 \cdot 10^{-3} m^2/s$, $Q = 1.5 eV$).

A third type of kinetic law for solid-state compound-forming reactions is characterized by an extremely strong, almost threshold-like temperature dependence. These reactions are limited by nucleation rather than growth rates. Transformation starts with random grains or islands of the compound phase, which subsequently grow together. Such behavior is found in several

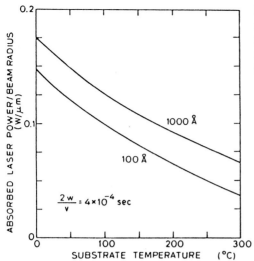

Fig.3.10. Combinations of absorbed laser power and substrate temperature required to form 10 and 100 nm thick films of Pd$_2$Si by partially overlapping laser scans [3.15]

rare-earth/Si systems, as well as in many compounds that are not first-growing phases, like Pdsi. Producing uniform layers by furnace annealing tends to be difficult in these cases. Scanning laser annealing has been demonstrated to produce uniform compound layers also in the case of Pdsi, but at a larger beam power than required for the first growing Pd$_2$Si phase [3.43].

Whatever the particular kinetic law, compound layers formed by a scanning-laser beam in solid-state reactions tend to be smooth and of uniform microstructure. Typically they resemble furnace-grown layers rather closely and usually consist of one single compound phase, in contrast to pulse-laser formed ones, as discussed in Chap. 4 (there are exceptions, like Pt, which forms a mixture of silicides during scanned laser annealing). Homogeneous silicide layers can even be obtained with refractory metals like Mo, W and Nb which require high reaction temperatures. Repeated scans may be needed, but heating the whole substrate, as in furnace annealing, is avoided. The localized heating achieved by scanned-laser annealing also causes much faster cooling of the processed material than conventional heat treatments. The formation of metastable compounds, not available through furnace annealing, has been reported in various metal-metal and metal-semiconductor systems after laser scanning [3.15]. Rapid cooling of the material after irradiation enables high-temperature phases to persist at room temperature without transformation to the thermodynamically stable low-temperature phase (this effect is also what is aimed at in hardening, see Sect. 3.2.5).

In addition to silicide growth, laser-induced compound synthesis is currently being explored in a number of innovative approaches. It is not always clear, and in fact sometimes irrelevant, whether these processes occur as pure solid-state reactions or involve melting and solidification. While it is difficult to give a well-seasoned assessment of these approaches at the present time, it is certainly worth mentioning some of these new developments.

Compound-Semiconductor Synthesis. Pulsed or CW irradiation of multilayer thin films has been used to obtain crystalline compounds in several III-V, II-VI and IV-VI systems [3.44]. Here, the average composition of the multilayer is chosen to coincide with that of the desired compound. The compounds are reported to be generally single-phase and stoichiometric. Although the actual reaction temperature has not been determined, it seems that the effect can be understood in terms of melting, mixing of the elements and subsequent solidification, a sequence we shall discuss in a broader context in Chap. 4.

Oxide and Nitride Formation. Irradiation of easily oxidizing materials like Si, Ti or V in oxygen-containing ambients results in strongly enhanced rates of oxide growth. Experiments were done with both CW [3.45] and pulsed [3.46] lasers at wavelengths ranging from infrared to ultraviolet. To what extent non-thermal (photon-induced) and nonequilibrium processes are responsible for the enhanced oxidation rates is still subject to debate [3.47]. The oxides are often found to be disordered or amorphous and – e.g., in the case of Ti – multiphase and nonstoichiometric.

Nitrides and oxinitrides of reactive metals can also be obtained from irradiation with pulsed or rapidly scanned CW beams in air [3.48]. The dissolution of oxygen into the metal upon rapid heating apparently causes a local and transient oxygen depletion near the surface which gives nitrogen a chance to react. More controllable nitride formation of semiconductor (Si, Ge) or metal (Fe, Ti, Zr, Hf, ...) substrates results if the irradiation is done in pure ammonia or nitrogen atmospheres [3.49].

Pyrolytic Film Decomposition. Irradiation of metalloorganic films by a scanned laser beam results in local thermal decomposition, as a result of which a metallic track is left in the beam path [3.50]. The heat of reaction (mainly due to exothermic pyrolysis of the organic carrier) appears to play a role similar to that of the heat of crystallization in explosive recrystallization (Sect. 3.2.2) and can give rise to periodic growth patterns reminiscent of those shown in Fig. 3.9.

Material Deposition by Gas or Liquid Phase Reactions. Laser beams are being used to achieve deposition of metallic or semiconducting materials by pyrolitic or photolytic decomposition of gaseous carrier media such as metallorganics, silane, etc. The beam is propagated through the carrier gas and a substrate is placed either parallel close to the beam or, more often, directly into the beam path. Decomposition of gas molecules, by direct absorption of photons, by impact on the laser-heated substrate or by thermal electrons from the latter, results in formation of chemically active species which then deposit on the substrate. The details of this process are still far from being completely understood. In some experiments liquid carriers have been employed. Here the reaction cell typically resembles an electrochemical cell in which one electrode is irradiated by the laser beam. Laser-enhanced reaction rates (with or without an applied voltage) seem to be largely due to electrode heating and/or improved transport in the liquid by thermal convection. For detailed discussions on laser-induced liquid or gas-phase reactions consult the reviews [3.51].

Pulsed-Laser Deposition. In this approach (which strictly does not even belong here) the laser beam is not directed at the material being synthesized, but serves as a kind of evaporation gun: A target material is vaporized and partially ionized by energetic ns to ms laser pulses in a vacuum chamber. Some of the vapor condenses on a substrate placed near the beam-impingement area. We shall discuss pulsed-laser deposition in some detail at the end of Chap. 5.

Novel and exotic techniques like these ensure, if nothing else, that the interest of materials scientists in laser beams is not going to fade away soon.

3.2.5 Transformation Hardening

It has been known since old ages that metals can be hardened by first heating them to some transformation temperature and then quenching them rapidly, e.g., by a plunge into cold water. That laser beams have been found suitable for the job is not surprising given their ability to provide rapid and controllable heating and subsequent cooling of a metal surface.

The physical mechanisms responsible for the hardening effect of a heating and cooling cycle can be quite complex and depend on the composition and microstructure of the metal to be hardened. Conventional hardening of carbon steel requires heating to a temperature above the α-γ transition temperature (between 750° and 900°C, depending on the carbon content) where the soft pearlite phase transforms into austenite and carbon particles dissolve. Upon subsequent rapid cooling the austenite transforms into metastable martensite, whereas excess carbon forms various kinds of precipitates which contribute importantly to the mechanical properties of the hardened material. The minimum cooling rate to avoid formation of unwanted soft modifications is of the order of 10^3 K/s in carbon steel. Even more complex processes are involved in hardening of alloyed steels. A role is also played by thermal stresses and hydrostatic pressure created during the thermal cycle. They do not only influence the phase equilibrium at high temperature but may also lead to plastic deformation and work hardening.

The standard lasers for hardening of ferrous alloys are CW CO_2 lasers with beam powers of up to several kW. Coatings to reduce metal reflection are often used. To achieve uniform hardening over extended areas the beam is often oscillated rapidly in a direction perpendicular to the scan direction, in order to produce the thermal equivalent of a line-shaped source [3.52]. Typical beam dwell times range from $0.01 \div 1s$. Hardenable depths turn out to be of the order of two or three mm under typical conditions. There is no evidence that scanned-laser beam hardening involves any unusual metallurgical processes. The main difference to conventional techniques is that heating and cooling tends to be faster and is concentrated near the surface. There is, on the other hand, also less time available for the material to reach equilibrium at high temperature. This is an undesirable feature, but it can largely be compensated for by using relatively homogeneous material with finely dispersed carbon to begin with. The rapid heating also limits the maximum material depth that can be hardened in a laser scan, because it favors surface melting and evaporation. Beneficial is that bulk heating and associated thermal distortions are avoided.

Modelling the process of laser hardening is readily done on the basis of the theory presented in Sect.3.1. The process parameters – laser power, spot size and scanning speed – are determined by the requirement that the volume to be hardened is heated to a temperature above the transformation temperature but below the melting point, and that the cooling rate is sufficiently large. Evaluation of (3.20) in the case of a scanned Gaussian beam, or of (3.19) for the case of irradiation of an extented area, gives all the necessary information. As an example, Fig.3.11 shows combinations of absorbed irradiance and interaction time (dwell time or pulse duration) for hardening of cast iron with a melting point of 1200°C [3.53]. Note that the heated zone becomes very narrow at high irradiance because the rapid surface heating limits the allowable interaction time. Deeper heating can be

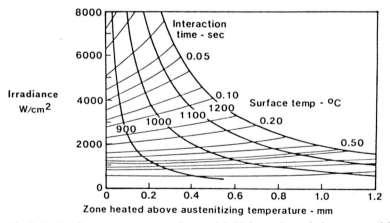

Fig.3.11. Surface temperature and material thickness heated above austenitizing temperature ($\simeq 800°C$) as a function of absorbed irradiance and interaction time, in laser hardening of cast iron (3.5% carbon, melting point 1200°C) [3.53] (Courtesy UTRC)

achieved with lower beam power and slower scanning, but at the cost of a smaller cooling rate. Insufficient cooling rates lead to partially tempered rather than hardened material. This again limits the practically achievable case depth to about 3 mm in steel.

4. Melting and Solidification

Arguably the most interesting material modifications by laser irradiation are those involving melting and solidification. Melts have high atomic mobilities and usually unlimited solubilities, enabling structures that are simply not available from pure solid-state reactions. Examples range form near-perfect crystals, intricately structured alloys and composites to glassy metals. The basic sequence that produces this variety of structures under transient laser irradiation is always the same: Melting of a surface layer and redistribution of atoms in the melt, followed by more or less rapid solidification. We shall summarize this basic sequence as *laser remelting*. There are two parameters which largely determine the structure resulting from a laser remelting process: the composition of the melt and the velocity of the solid-liquid interface.

The experimental work available to date on laser remelting can roughly be divided into three categories: (i) *regrowth of ion-implanted substrates*, (ii) *surface alloying*, and (iii) *melt quenching*.

The wide-spread interest in the first category of activities can be traced back to the pioneering work of Russian scientists [4.1] who observed that irradiating ion-implanted semiconductors with Q-switch laser pulses could restore crystalline order far more perfectly than furnace (solid-state) annealing – apparently because crystal growth occurred from a melt, rather that directly from the amorphous phase. However, the technique has turned out to be more than just another implementation of the old art of growing a crystal from its melt. Solidification velocities observed for short pulses are measured in m/s, ten orders of magnitude or so faster than in conventional liquid-phase epitaxy, and work in this regime has produced – if not actual applications in the semiconductor industry – at least important new insights into the ways crystals grow.

Whereas ion-implantation is limited to a few at.% of admixture to a substrate material, *surface alloying* enables mixing of two (or more) elements at any desired composition. That the alloys tend to be nonuniform in their microstructure does not impair their basic appeal, which is to allow the surface properties of a material to be made independent of its bulk properties. Much of the earlier work in alloying has been concerned with semiconductor substrates, whereas recent emphasis has shifted more towards metallic substrates. Here an additional motivation is the considerable saving in expensive or scarce alloying elements achieved if bulk alloying is replaced by surface alloying.

The application of lasers in *melt quenching* with the aim of obtaining glassy alloys is the most recent of the three groups of laser remelting procedures, although the potential was realized long ago [4.2]. Surface remelting with sub-μs pulses can yield cooling rates of 10^{10} K/s or more, many orders of magnitude faster than conventional mechanical methods like melt spinning. As a result, many materials are becoming available in glassy form that were previously not thought to be glass-formers.

It should not be concealed that laser processing in the melt regime, for all its appeal, also has its shortcomings – at least from an applications point of view. Beam powers are much higher than those typical in solid-state processing, particularly in experiments with short pulses. Self-induced coupling phenomena, as discussed in Chap. 2, tend to make controlled heating more difficult. As a result, the "power-window" to conduct melting experiments without material damage can be quite narrow. In addition, the mechanical forces that arise during rapid melting can lead to destruction of delicate samples like semiconductor structures or thin films, or produce intolerable surface warping. Accurate control of processing parameters is therefore critical for applications in the regime under discussion – more so than in all others considered in this volume.

We start this chapter with a section discussing models and fundamental aspects of laser remelting, with emphasis on its two key variables, the melt composition and the interface velocity. Subsequent sections then focus on the three approaches to laser remelting defined above.

4.1 Fundamentals

4.1.1 Regimes of Laser Remelting

There are two fundamental aspects to be considered in the sequence we call *laser remelting*. The first is chemistry – in order to understand a remelted structure we need to know the chemical properties of the elements concerned as well as their concentration. Some of this information is given – for the case of thermodynamic equilibrium – by the phase diagram. This is not sufficient, however, since equilibrium is seldom achieved in a process as rapid as laser remelting. The second aspect, then, is kinetics, which deals with the paths and velocities at which phases form and equilibrium is approached. Thermodynamics and kinetics are the topics of Sects. 4.1.3-5 (we shall limit the discussion to binary systems, because they are simpler and because only little quantitative work has been done on more complex systems).

In order to make efficient use of thermodynamic and kinetic information, we first need a model description of the laser remelting process, which quantifies the flow of particles and heat in the irradiated zone. Such modelling takes as its input the initial elemental distribution of the sample as well as the parameters of the heat source and yields, as its main results, the transient melt-concentration distribution and the velocity of the melt-solid interface. Let us first consider particle flow. To keep the mathematics simple, we shall consider one-dimensional geometries only.

The initial configuration of the sample is characterized by some elemental distribution as a function of depth beneath the surface. Concentration profiles of three types of samples often used in laser remelting are displayed in Fig.4.1:

(a) The Gaussian function, describing approximately the profile obtained by ion implantation of an "impurity" atom B into an elemental substrate A.
(b) The rectangular distribution, occurring in the case of a deposited layer B on top of a substrate A.
(c) The square-wave profile, describing a multilayer of alternating deposited films of B and A on top of some inert substrate.

Upon irrdiation, the material heats up and melts, and the elemental distribution starts to change by diffusion. Diffusion in the short premelting period can generally be neglected since the diffusivities of most materials increase by several orders of magnitude upon melting. Diffusivities in the range 10^{-5} to 10^{-4} cm^2/s are typical in molten metals, and changes in the original elemental distribution within the irradiated structure occur even if the melt lifetime is only a fraction of one μs. The effect of diffusion for a time t on the elemental distribution is also shown in Fig.4.1 (shaded).

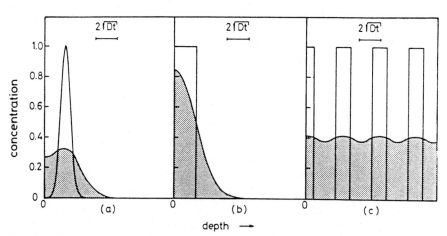

Fig.4.1. Concentration profiles before and after diffusion for a time t: (a) Gaussian profile (ion-implanted single-crystal), (b) rectangular profile (deposited film on a substrate), (c) square-wave profile (multilayer)

Mathematically, particle diffusion is similar to diffusion of heat, except that there are now several – at least two – diffusing particle species, which differ in structure and mass and which need not have the same "intrinsic" diffusivities. The simplest case, which we assume here, is that the total volume occupied by the elements is not changed by diffusion. Interdiffusion of two elements can then be described in terms of a single "mutual" diffusivity D, which may depend on the composition. The differential equation for isothermal diffusion (known as Fick's second law) can be written as

$$\frac{\partial X}{\partial t} = \nabla(D \nabla X) .$$ (4.1)

If D is taken as a constant, (4.1) can be solved analytically by methods similar to those used in Sect.3.1 [4.3]. In particular, instantaneous concentration profiles after diffusion for a time t in a semi-infinite melt (z > 0) are obtained from

$$X(z,t) = \frac{1}{2\sqrt{\pi Dt}} \int_0^\infty X_0(z') \left[\exp\left(-\frac{(z-z')^2}{4Dt}\right) + \exp\left(-\frac{(z+z')^2}{4Dt}\right) \right] dz'$$ (4.2)

where $X_0(z)$ is the initial distribution. This equation (which was used to construct Fig.4.1) neglects the fact that the melt lifetime is not everywhere the same. For more accurate calculations one may allow t to vary within the melt according to the finite velocity of the solid-melt interface (Sect.4.1.2) [4.4]. Alternatively, (4.1), along with the appropriate initial and boundary conditions, can be solved numerically by the finite-difference technique. The procedure is completely analogous to the one outlined for the heat flow equation in Appendix C.

Figure 4.1 demonstrates that interdiffusion for a time t produces mixing of the elements A and B over a characteristic length $2\sqrt{Dt}$. The resulting distribution (shaded in the figure) can be characterized as inhomogeneous and diluted in case (a), inhomogeneous and concentrated in case (b), and concentrated but homogeneous in case (c). It should be no surprise that solidification in the three cases produces quite different structures. The situation may, somewhat simplified, be summarized by saying that

case (a) leads to epitaxial growth of A,
case (b) to nonuniform and heterogeneous A-B alloy formation, and
case (c) to uniform and often metastable phase formation.

In Sects.4.2-4 we shall consider the three regimes corresponding to the three basic configurations of Fig.4.1 in some detail. Let us now turn to heat flow.

4.1.2 Heat Flow and Latent Heat

For a crystal to grow, it (or, more accurately, the crystal-melt interface) must be kept at a temperature below the equilibrium melting point T_{sl}. Since crystal growth liberates latent heat, this extra heat must be continuously carried away by heat conduction to preserve the undercooling. Similarly, heat must constantly be furnished to the interface to preserve superheating during melting. This requirement determines the velocity of the interface. In the following we neglect the elemental inhomogeneity of the material (it could approximately be accounted for by using suitably averaged values for the thermophysical constants).

A planar interface moving at the velocity u can be thought of as a moving source or sink of heat, creating a heat flux equal to $-u\Delta H_{sl}/V$, where ΔH_{sl} is the latent heat and V the molar volume (taking u positive for solidification and negative for melting). Motion of the interface is sustained by the local heat flow provided that

$$\frac{u\Delta H_{sl}}{V} = K_s \left.\frac{\partial T_s}{\partial z}\right|_{z_i} - K_l \left.\frac{\partial T_l}{\partial z}\right|_{z_i} \tag{4.3}$$

where K_s and K_l are the conductivities of the solid and the liquid, respectively, and z_i is the coordinate of the interface. The velocity is thus proportional to the difference between the temperature gradients on the liquid and the solid "faces" of the interface. The temperature of the interface, in turn, is not determined by heat flow but by kinetics, as we shall discuss in Sect. 4.1.4. A slowly moving interface (driven by a small thermal gradient) will be at a temperature close to T_{sl}, while for faster movement the temperature will be appreciably higher or lower. During very rapid cooling the "heat flow" velocity may even exceed the maximum possible kinetic velocity. In this case the melt will not crystallize but freeze into a glass.

To discuss the heat-flow aspects of melting and solidification at ordinary velocities it is convenient to ignore kinetics, and to assume that the interface is at a fixed temperature, usually taken to be the equilibrium melting point. The interface is then an isothermal surface, the motion of which is obtained by solving the appropriate heat-flow problem.

The simplest approach is, obviously, to neglect the latent heat. The heat-flow problem then reduces to that treated in Sect. 3.1. In one-dimensional heat flow the isothermal surface $T = T_j$ is a plane with coordinates z_j, t_j given by $T(z_j, t_j) = T_j$. To illustrate the concept, Fig. 4.2 depicts the positions of various isothermal planes for the normalized distribution (3.26). The isothermal planes (and hence the interface) move into the solid during, and shortly after, the laser pulse; thereafter they return to the surface. The velocities are related to the temperature by $u = (\partial T/\partial t)/(\partial T/\partial z)$ and are seen to become infinite in the end. This is a consequence of our

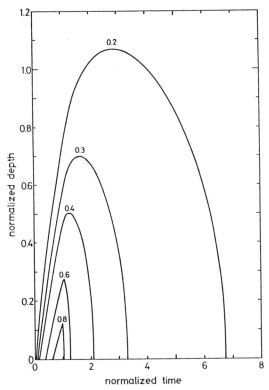

Fig.4.2. Positions z^* of various isothermal surfaces as a function of time t^*, calculated from (3.26). The parameter gives the value of the normalized temperature θ^*

assumption of a thermally insulated surface where the temperature gradient vanishes by definition.

If latent heat is included, solution of the heat-flow equation becomes considerably more difficult. Analytical treatments have yielded solutions only for a few special cases not of much interest for our problem [4.5], and one has to resort to numerical methods. The main difference between the present heat-flow problem and that without a phase change is that now the state of the material is no longer uniquely defined by the temperature, since near T_{sl} it may be either solid or liquid. In the following we shall, for simplicity, take the molar volume V to be constant and the interface temperature to equal T_{sl}. A convenient quantity that uniquely defines the state of the material is the difference in total enthalpy between the ambient temperature and temperature T

$$\Delta H(T) = \int_{T_0}^{T} c_p (T')dT' + [T > T_{sl}]\Delta H_{sl} .\tag{4.4}$$

The symbol $[T > T_{sl}]$ equals one above and zero below the chosen transition temperature. The use of a step function here implies that we assume the kinetics to be "infinitely fast", i.e., the material instantly switches its state at a fixed temperature, no matter what the interface velocity may be. This assumption has implicitly been made in most calculations of interface velocity published to date [4.6-8] and we adopt it for this discussion (for calculations that allow for finite transformation kinetics, see [4.9]). Proceeding then, we can write the heat-flow equation as (retaining heat flow along the z-axis only)

$$\frac{\partial \Delta H}{\partial t} = V \left[\frac{\partial}{\partial z} \left(K \frac{\partial T}{\partial z} \right) + J(z, t) \right] \tag{4.5}$$

where J is the power density of the heat source. Eq.(4.5), together with (4.4), may be regarded as a generalisation of (3.6), valid for arbitrary temperatures and suitable for numerical evaluation. A simple numerical method of solution is outlined in Appendix A.3.

Let us now consider a laser remelting cycle as modelled with the numerical method. Temperature distributions and interface motion turn out to be deeply influenced by the turnover of latent heat. This is illustrated by Figs.4.3,4 for the case of a semi-infinite solid, irradiated by a rectangular pulse. Figure 4.3 displays temperature profiles at various times during and after irradiation, while Fig.4.4 gives the positions of the interface during the melting and solidification cycle for various values of the latent heat. Surface absorption and constant thermal parameters were chosen for this calculation to allow a comparison with Fig.4.2 for the case without phase change; the same normalized coordinates are used in the graphs. The calculation demonstrates that the latent heat has a rather strong impact on the interface velocity during cooling. A latent heat of only 1/6 of the heat content of the solid at the melting point already doubles the lifetime of the melt (Table A.4 gives actual values of the ratio of latent heat to the heat content of the solid at T_{sl}). Furthermore there is, perhaps contrary to intuition, virtually no temperature gradient in the melt during solidification. This means that the second term on the RHS of (4.3) is negligible compared to the first one, and the solidification velocity, for a given latent heat, is determined by the conductivity of the solid phase alone. This fact has been used to influence the solidification velocity in Si, where the conductivity varies strongly with temperature (Fig.3.4), by varying the substrate temperature [4.10]

The solidification velocity obviously depends on the fluence and the duration of the heat pulse, too. A more energetic or longer pulse results in a larger melt depth, and since the stored heat is larger the thermal gradients become shallower. In strongly absorbing media and for not too short pulses $(1/\alpha \ll \delta)$ the thermal gradient near the surface is roughly $T(0)/\delta \propto t^{-1/2}$ (Fig.3.1) and the velocity should scale with the pulse duration like $t_p^{-1/2}$.

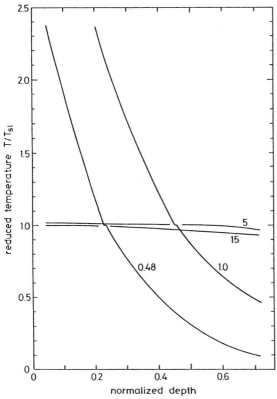

Fig.4.3. Numerically calculated temperature profiles at various times during melting and solidification. Time and distance are expressed in terms of the same normalized coordinates as used in Fig.4.2

This is borne out in Fig.4.5, which illustrates the impact of the substrate temperature (i.e., K_s) as well as the pulse duration on the interface velocity in crystalline Si irradiated with pulses in the nanosecond regime [4.7]. The basic reliability of the numerically calculated interface velocities in Si has been verified by direct measurements in which the transient conductance of suitably shaped samples under laser irradiation was monitored [4.11]. These measurements confirmed, in particular, that very large crystal-growth velocities, of the order of several m/s, are reached in materials irradiated by ns laser pulses.

The same type of calculation can, of course, be applied to melting by scanned CW beams, except that cylindrical or three-dimensional Cartesian coordinates must be used. Figure 4.6 shows calculated melt depths in Si scanned by an Ar laser beam for a range of beam powers and scanning speeds [4.12]. Note that the sensitivity of the melt depth to the laser power decreases towards larger scanning speeds, suggesting that melting should be easier to control by scanning an intenser beam more rapidly.

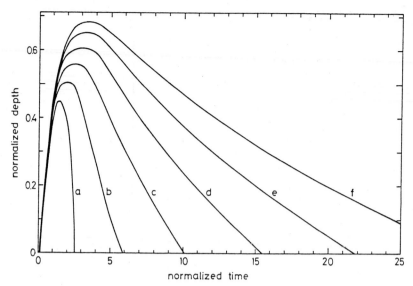

Fig.4.4. Position of the l-s interface as a function of time for various ratios of the latent heat to the enthalpy at the melting point, $\Delta H_{sl}/\Delta H(T_{sl})$. For curves a-f this ratio is 1/3000, 1/6, 1/3, 1/2, 2/3 and 5/6, respectively. Curve e corresponds to Fig.4.3. The pulse energy is kept constant

Having gained some insight into the macroscopic mass and heat flow during laser remelting, let us now consider how this information relates to the microscopic processes that determine the final structure.

4.1.3 Thermodynamics

Thermodynamic equilibrium between a crystal and its melt requires that both phases have equal free energies. The Gibbs free energy of a phase at absolute temperature T is defined as G = H–TS, H and S being the enthalpy and entropy, respectively. The equilibrium melting point T_{sl} is that temperature at which the liquid and the solid have equal free energies at 1 atmosphere. Any deviation of temperature from T_{sl} causes one phase to become unstable, i.e., to have a higher free energy with respect to the other. The difference in free energy between the phases

$$\Delta G_{sl} = G_s - G_l = \Delta H_{sl} - T\Delta S_{sl} \simeq \Delta H_{sl}\Delta T/Tsl \qquad (4.6)$$

is the chemical driving force for promotion of the interface. Here, ΔH_{sl} (the latent heat) and ΔS_{sl} are the enthalpy and entropy differences between the solid and the liquid, and $\Delta T \equiv T_{sl}-T$ is the *undercooling*. The substitution $\Delta S_{sl} = \Delta H_{sl}/T_{sl}$ made above strictly holds only a equilibrium, but is

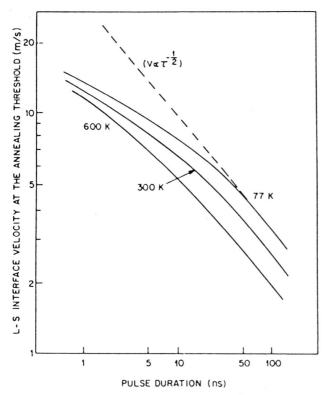

Fig.4.5. Calculated interface velocity in Si (crystalline substrate with a 100 nm thick amorphous surface layer), irradiated by ruby laser pulses, as a function of pulse duration for various substrate temperatures [4.7]

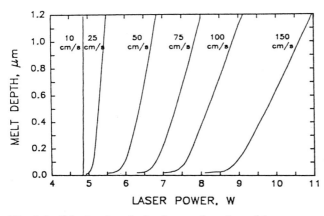

Fig.4.6. Calculated melt depth as a function of laser power in single-crystalline Si scanned at different velocities [4.12]

used as an approximation for $T \neq T_{sl}$, too. Note that with the sign convention of (4.6) we have $\Delta G_{sl} < 0$ in the undercooled melt.

Equation (4.6) holds for elemental as well as for mixed phases, but the free energy of a mixture at a given temperature is a function of composition. Considering a binary mixture A–B, where X denotes the molar fraction of element B, we have

$$G(X) = (1-X)G^A + XG^B + G_M = (1-X)g^A + Xg^B . \qquad (4.7)$$

Here, G^A and G^B are the free energies of the pure elements A and B, G_M is the free energy of mixing (or formation), and g^A and g^B are the partial molar free energies (or chemical potentials) of the components in the mixture. A partial molar quantitity is the increase in the respective total quantity of the mixture upon addition of an infinitesimal amount of pure substance (per mole added). The total and partial free energies are related by

$$G = g^A + X(\delta G / \delta X) = g^B - (1-X)(\delta G / \delta X) . \qquad (4.8)$$

This means that the chemical potentials at some composition X_1 are found by drawing the tangent to the $G(X)$ curve and extending it to the respective elemental axes. A schematic free energy curve of a particular phase in a binary mixture is shown in Fig.4.7 (the detailed shape of the curve depends on its free energy of mixing).

Equilibrium between two phases requires that, in addition to pressure and temperature, the chemical potentials of all components are the same in both phases. Figure 4.8 sketches free-energy curves for a liquid and two solid phases at the temperature T_1 below the melting points of both solids. The solid compositions X_{s1} and X_{s2} in equilibrium with each other are obtained by drawing the common tangent to the respective curves. The driving force for solidification at the temperature T_1 of a melt of composition X is

$$\Delta G_{sl}(X) = (1-X)\Delta g_{sl}{}^A + X\Delta g_{sl}{}^B \qquad (4.9)$$

and can be read off the figure. $\Delta G_{sl}{}^{(1)}$ and $\Delta G_{sl}{}^{(2)}$ are the driving forces for the initial precipitation of solids 1 and 2, respectively, while ΔG_{sl} is the driving force for the formation of the equilibrium configuration which is a mixture of both solids. Note that, as a rule, the driving force for any single solid phase decreases with increasing compositional difference between it and the melt.

The above concepts are concepts of equilibrium thermodynamics, and the reader may perhaps wonder about their relevance in laser processing, where large thermal gradients and irreversible flow of heat tend to create *a state of strong nonequilibrium*: Nevertheless, a solid-liquid interface can be a T_{sl} – and hence at rest – even while heat is flowing through it, so the

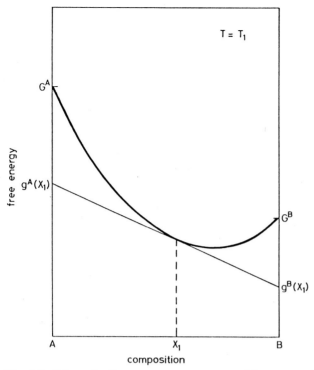

Fig.4.7. Schematic free energy-versus-composition curve. G^A and G^B are the free energies of pure elements A and B at temperature T_1, g^A and g^B are the chemical potentials at composition X_1

phase there must be in some sort of equilibrium. This is called *local equilibrium*. While the heat flow by itself is irreversible, the entropy it produces is generally not available for the phase transition, a local phenomenon taking place at a narrow interface [4.13]. Equilibrium in this local, restricted sense must be understood whenever equilibrium concepts are applied to a transient process. Experience shows that most laser-induced material phenomena, even for pulses in the ns range, can well be understood from local-equilibrium thermodynamics. Exceptions are the phenomena of impurity trapping and glass formation, to be treated in Sects.4.2.2 and 4.4.

Systematic application of the *common tangent rule* yields the equilibrium phase diagram. Even when global (in contrast to local) equilibrium is never reached in an alloying process, the phase diagram can provide important clues about the phase to be expected. While phase diagrams come in greatly varied shapes and forms, we may distinguish four main types which exhibit typical behavior upon melting and solidification. The four types, roughly in sequence of progressively more positive heat of mixing, can be characterized as follows (examples are listed in Table A.5):

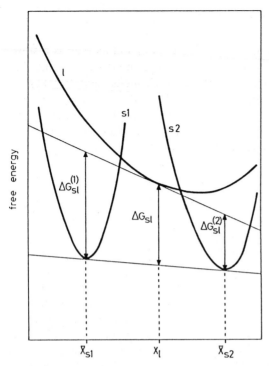

Fig.4.8. Schematic free energy curves of a liquid and two solids at a temperature below the melting points of the latter. $\overline{X}_{s1}$ and $\overline{X}_{s2}$ are solid compositions in equilibrium with each other. Also shown are the driving forces for initial precipitation of solids 1 and 2 from a melt of concentration X_l, as well as for solidification of the latter into the equilibrium configuration

(1) Partial or complete solid solution formers,
(2) compound formers,
(3) systems with immiscible solids (simple-eutectics), and
(4) systems with immicible melts.

The last group of systems is rather unfavorable for alloying by melting and solidification, whereas the others differ mainly in the microstructure of the alloys they form. Type-1 systems produce homogeneous (single-phase) alloys at any composition, while Type-2 and Type-3 systems usually produce heterogeneous grained alloys with grain sizes inversely related to the cooling rate. The basic behavior of a system, as predicted by the phase diagram, is often modified by the presence of metastable phases, particularly if cooling rates are high and the resolidified structure never comes close to global thermodynamic equilibrium.

So much for thermodynamics. The question about the actual velocity of phase formation is not answered by thermodynamics, but requires a microscopic model of the phase transition.

4.1.4 Interface Kinetics

The kinetic picture of equilibrium between a crystal and its melt is a situation in which the same number of atoms cross the interface per unit area and unit time in either direction, i.e., the crystal "melts" and "freezes" at the same rate. Both rates are assumed to be thermally activated, with the activation energy for freezing being larger than that for melting by the latent heat per particle. Net movement of the l-s interface results from an imbalance of the two rates. The velocity of a plane interface can be expressed as

$$u = aA\nu_j \left[1 - \exp\left(\frac{\Delta G_{sl}}{RT}\right) \right] . \tag{4.10}$$

Here a is the interatomic spacing in the growth direction, A is the accommodation probability for atoms in the growing phase, and ν_j is a thermally activated jump frequency. The accommodation probability for growth can be interpreted as the fraction of sites at the crystal boundary, which are next to a step or ledge, where attachment of a new atom is energetically favorable. For most metals A is close to unity at all temperatures, while for semiconductors, halogens and metalloids at small undercooling it is small and anisotropic – larger for non-closely packed interface directions than for closely packed ones, making growth along a $\langle 100 \rangle$ direction in a cubic crystal faster than along $\langle 111 \rangle$ [4.14]. Figure 4.9 exhibits a schematic of u as a function of the interface temperature (for A = 1). Note that for small undercooling ($\Delta G_{sl} \ll RT$) the velocity is proportional to the undercooling. Since in laser remelting the interface velocity is dictated by heat flow, see (4.3), the actual undercooling at the interface can be estimated by equating the "kinetic" and the "heat-flow" velocities. For very large undercooling the bracketed term in (4.10) tends towards unity and the velocity becomes small and proportional to the thermally activated jump frequency. Heat flow then plays no role in determining the interface velocity. This is the regime of thermally activated growth discussed in Sect. 3.2.

The activation energy determining ν_j can be estimated from the activation energy for self-diffusion or for viscous flow in the melt. In pure metals this is small, and atoms seem to remain relatively mobile down to temperatures far below ambient. *Turnbull* and his associates have suggested that growth in pure metals is not limited by mobility but only by the impingment rate on the interface [4.15]. An upper limit for ν_j is

$$\nu_j < u_s/a \tag{4.11}$$

where u_s is the sound velocity in the melt; maximum growth velocities would thus be expected to be in the 10^3 m/s range.

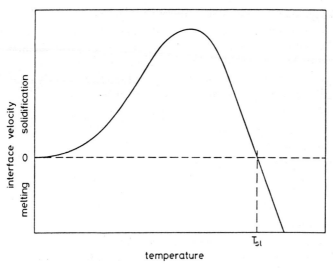

Fig. 4.9. Schematic variation of the liquid-solid interface velocity with temperature

In solidification of molecular or covalently bound materials like SiO_2 (as well as in solid-state crystallization of amorphous Si or Ge) atomic rearrangements are sluggish even at elevated temperature. Crystallization here requires coordinated motion of whole groups of atoms, referred to as activated complexes, rather than just jumps of single atoms. The corresponding rearrangement rate has been expressed as [4.15]

$$\nu_j = n\nu_{fo}\, e^{-\Delta G'/RT} \tag{4.12}$$

where $\Delta G'$ is the free energy of formation of the activated complex, and n is the number of individual rearrangements resulting from a single activation. Since pairs of dangling half-bonds, once formed, may migrate and enable several atoms to regroup before they recombine, n may be larger than unity. Intermediate cases are materials like Si and Ge which form covalent crystals but metallic melts. The change in atomic coordination as well as the 10% change in density required for crystallization make l-s interface motion slower than in pure metals, but it still remains much faster than in nonmetallic melts. Growth velocities for a given undercooling depend on the crystal orientation, and are largest for $\langle 100 \rangle$ and smallest for $\langle 111 \rangle$ growth, due to differences in the accommodation probability [4.16]. Si(100) is observed to grow epitaxially up to a velocity around 20 m/s, but it turns amorphous at larger velocities. We will come back to this phenomenon in Sect. 4.2.

Growth of a compound crystal from a binary melt is more complicated, particularly if the crystal and the melt have different compositions. *Jackson* [4.17], considering growth of a solid solution, applied reaction-rate theory to both species independently, with each component growing at a velocity determined by its own chemical potential change and proportional to its concentration in the melt. The total interface velocity is then just the sum of the component velocities, weighted with their concentrations in the crystal. However, growth of an ordered compound crystal requires strictly coordinated, rather than independent, component growth processes. The usual assumption is that the general form of (4.10) (with a suitable driving force) is still valid. The necessary redistribution of melt atoms at the interface can be allowed for in an approximate way by replacing the jump frequency by the frequency for diffusive transport in the melt

$$\nu_j \simeq D/a^2 \qquad\qquad\qquad (4.13)$$

where D is the melt diffusivity at the interface. For metallic compounds ν_j would be of the order of 10^{11} s^{-1}, significantly smaller than for pure metals. If the melt composition differs from the composition of the crystal, growth is further limited by long-range mass transport, which is not accounted for by (4.13). These limitations may explain why metallic glass formation by melt quenching (Sect.4.4) is achieved for alloys but not for pure metals.

4.1.5 Nucleation

The situation discussed so far is that of a crystal in contact with its melt, as it occurs in a surface-melted semiconductor wafer. If only one phase is present initially, then the other must, under normal circumstances, be nucleated before any transition can occur. Nucleation involves the formation of a new surface between the phases, which is energetically expensive and occurs only if the driving force exceeds a certain critical value.

How, and when, does a crystal melt under laser irradiation? The question is not as pointless as it may sound at first. Initiation of melting with moderate heating rates is a nucleation and growth process. Melt nucleation in solids internally heated by light absorption has been observed long ago [4.18]. The amount of superheating required is largest for materials with highly viscous melts, such as quartz. However, significant superheating may be expected in any material at the prodigious heating rates available with short laser pulses (10^{15} K/s or more), although the question has attracted little attention so far. There appears to be a limit for the superheating of solids, given by the criterion that the resistance of the crystal to shear vanishes, causing it to melt continuously, without prior nucleation. Based on *Frenkel's* [4.19] theory of melting, *Martyniuk* [4.20] predicted the limiting relative superheating $-\Delta T/T_{sl}$ in electrically exploded metal wires to be between 15 and 30%. Even larger superheating has recently been reported in

GaAs irradiated by 150 fs pulses, where melting was concluded to occur between 500 and 1000 K above the equilibrium melting point [4.21].

Crystal nucleation, in contrast to melt nucleation, is of great practical significance in melt solidification. It is undesirable in laser annealing because it results in poly- rather than single-crystalline material.

The theory of homogeneous crystal nucleation is based on the idea that in an undercooled melt crystal-like clusters form by way of fluctuations. Growth of clusters is driven by the free energy of crystallization ΔG_{sl}, but opposed by the crystal-melt interface energy. Assuming the latter to be equal to the macroscopic interface tension σ, and taking the cluster to be spherical, the two opposing forces are found to balance each other once the cluster has reached a critical radius. The free energy of forming such a critical cluster, or nucleus, is

$$\Delta g_N = \frac{(16\pi/3)\sigma^3 V^2}{\Delta G_{sl}^2} . \qquad (4.14)$$

Once a nucleus has reached the critical size it no longer has to rely on fluctuations but can grow in the undercooled melt in the normal way. The steady-state rate of formation of nuclei is usually written as

$$Y(T) = A_N \, N\nu_j \, e^{-\Delta g_N /kT} \qquad (4.15)$$

where N is the melt particle density and A_N is a constant. The very strong temperature dependence of Y below T_{sl} leads to threshold-like behavior. The "threshold" temperature (referred to as the homogeneous nucleation temperature T_N) was estimated to be about $0.8T_{sl}$ in pure metals, with the corresponding critical nucleus containing about 200 atoms [4.18].

The nucleation rate (4.15) holds under isothermal conditions. In melts cooled rapidly from a temperature near T_{sl} to a temperature below T_N measurable nucleation is found to occur only after a finite delay or time lag. This may be interpreted as the time required to establish an equilibrium population of clusters corresponding to the new temperature [4.22, 23]. The time-lag can be expressed as

$$\tau_N \simeq \frac{RT\sigma V}{a\nu_j \Delta G_{sl}^2} \qquad (4.16)$$

and should be minimum roughly at the temperature where the steady-state nucleation rate is at its maximum. It obviously cannot be shorter than the number of atoms in a critical nucleus divided by the jump frequency. *Peercy* and coworkers [4.24], investigating pulsed laser annealing in Sb-implanted Al, estimated the time required for AlSb nucleation from molten Al to be between 5 and 25 ns. If this result is typical for alloys, it may help to explain the surprising ability of ns pulses for binary metallic glass formation (Sect. 4.4).

In many practical situations nucleation is found to occur at much smaller undercooling than predicted by (4.15), due to heterogeneous nucleation. In the presence of a substance which is wetted by the nucleating material, the cluster surface exposed to the liquid for a given cluster volume is decreased, and so is the critical nucleus size. Suitable substances include impurity particles in the melt as well as material boundaries, e.g., a film-substrate interface. Nucleation of melting usually occurs heterogeneously at the surface of the heated solid, rather than in the interior. On general grounds, impurity-mediated heterogeneous nucleation may be expected to become less important at large driving forces where homogeneous nuclei, once forming, quickly outnumber the limited supply of heterogeneous nucleation centers.

4.2 Regrowth of Ion-Implanted Substrates

Ion implantation produces shallow buried regions containing implanted atoms some tens to hundreds of nm beneath the surface. The distribution profile resembles a Gaussian with a depth (the mean ion range) and a width (the RMS deviation) determined by the ion kinetic energy and the host material. Maximum concentrations are typically in the at.% range, limited by surface sputtering. Somewhat similar elemental distributions have also been produced by deposition of ultrathin buried layers.

4.2.1 Semiconductor Substrates

Doping semiconductor wafers by ion implantation, while offering better control of the dose and the concentration profiles than the conventional diffusion technique, also causes extensive lattice damage. In Si room-temperature implants of ions heavier than Zn at doses exceeding about 10^{14} cm^{-2} result in amorphization of the implanted layer, while lighter ions, lower doses or higher implant temperatures create at least extended defects. Subsequent annealing is required to remove the damage and to bring the dopant atoms to electrically active lattice locations, but furnace annealing often requires the wafers to be subjected to temperatures as high as 1100°C or is even found unable to fully restore the lattice. As alternative techniques, rapid solid-state annealing by scanned beams (Sect. 3.2.3) as well as surface remelting by sub-μs laser or electron-beam pulses (often termed *pulsed annealing*) have therefore attracted enormous interest.

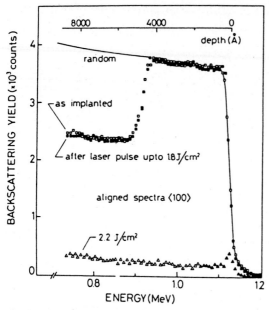

Fig.4.10. 2.0 MeV He$^+$ backscattering spectra of an ion-implanted Si(100) wafer with a 450 nm thick amorphous surface layer, before and after irradiation with 20 ns Ruby laser pulses of various fluences [4.25]

Lattice regrowth in pulsed remelting of implantated substrates has a sharp threshold, which depends on the thickness of the amorphous layer. Figure 4.10 exhibits MeV He$^+$ channeling spectra from a Si(100) wafer with a 450 nm thick amorphous layer, irradiated with 50 ns ruby laser pulses [4.25]. The lattice is restored to the quality of the virgin wafer at a fluence of 2.2 J/cm^2, while no improvement is seen for fluences up to 1.8 J/cm^2. The structure for fluences between about 0.5 and 1.8 J/cm^2 turns out to be polycrystalline, while for even lower fluences it is still amorphous. Equivalent observations are made also at other wavelengths, e.g., in the UV [4.26]. They are readily understood if one assumes the regrowth mechanism to be melting and subsequent liquid-phase epitaxy. For the amorphous layer to crystallize within the short interaction time it must be molten (the melt threshold for ruby-laser pulses is about 0.5 J/cm^2 in amorphous Si), but for it to regrow epitaxially the melt front must have penetrated into the single-crystalline matrix beyond the damaged layer. The regrowth threshold also depends on the structure of the damaged layer. Only slightly damaged or polycrystalline layers have higher thresholds than amorphous ones, because of their smaller absorption coefficient (Table A.1) and their larger heat capacity.

Pulsed laser remelting under the proper conditions produces regrown layers free of extended defects in both Si(100) and (111), implanted with the common Group-III and-V dopants. However, fluences sufficient for complete regrowth in (100) cut Si wafers yield heavily twinned material in Si(111), where only a somewhat larger fluence, other parameters being equal, gives a comparable crystal quality [4.27]. The reason for this puzzling discrepancy turns out to be directly related to the interface velocity [4.28, 29]. Interface velocities in Si for pulse durations around 50 ns are of the order of several m/s; and they decrease somewhat with increasing fluence (Sect.4.1.2). It must be concluded that these velocities are close to the limit at which defect-free Si crystals can grow, and the limit is apparently lower for growth along the ⟨111⟩ than along the ⟨100⟩ direction. This trend can be ascribed to the anisotropy of the accommodation coefficient A, as discussed in Sect.4.1.4. The limiting velocity for defect-free growth in Si (111) has been determined to be near 6 m/s [4.28]. In (100) material defect-free growth is observed up to a velocity of 18 m/s, while for an even larger interface velocity the nature of the transition alters dramatically – the Si turns amorphous. The same happens in Si(111) at only 14 m/s.

The production of amorphous Si by rapid cooling of the melt was first demonstrated in 1979 [4.30, 31]. The short pulses used to melt the surface of single-crystal wafers to a depth of only a few tens of nm left behind amorphous patches distinguished from their crystalline surroundings by a somewhat higher reflectance. The reason why Si turns amorphous is obviously that cooling is too rapid for the crystal to grow. However, the situation is different from that of glass formation by melt quenching, to be discussed in Sect.4.4. Amorphous Si is a four-fold coordinated semiconductor while molten Si is a metal with a coordination number near 11. Whereas glass formation can be thought of as a mere freezing of the melt, the transformation from liquid to amorphous Si seems to require a first-order phase transition [4.15, 32].

Figure 4.11 represents a free-energy diagram of Si (showing free-energy differences with respect to the crystalline phase), where amorphous Si (a) is treated as an independent phase with an equilibirium melting point T_{al} [4.33]. The latent heat of the l-a transition was reported to be 37 kJ/mol, as compared to 50.3 kJ/mol for the l-c transition. Consider what happens if the melt (l) is cooled below the equilibrium melting point of the crystal (T_{lc}). ΔG_{lc}, and hence the crystal growth velocity, increase, but below T_{al} nucleation and growth of amorphous Si competes with crystallization. That the amorphous rather than the crystalline phase (which has a still larger driving force) grows from the melt only indicates that the former is kinetically favored [4.34]. The apparent maximum crystal-growth velocity is thus no intrinsic limit but simply the value of the growth velocity at that undercooling at which the amorphous phase overtakes the crystalline one [4.35]. Melting of the amorphous phase, as observed in pulsed-laser remelting, can be understood from similar kinetic arguments: Once the amorphous phase is heated to a temperature between T_{al} and T_{cl}, melting occurs simply because it is faster than crystallization (melt-in velocities in Si of several 100 m/s

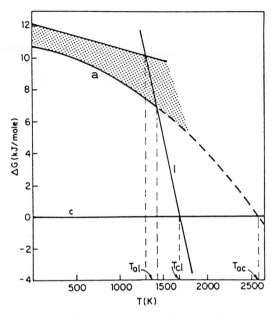

Fig.4.11. Free energy of amorphous (a) and liquid (ℓ) Si relative to that of crystalline (c) Si as a function of temperature. The dashed area indicates the uncertainty in the thermodynamic data for the amorphous phase [4.33]

during irradiation with ps pulses was reported [4.36]). On the other hand, kinetic hindrance of melt nucleation may explain the observation of solid-state crystallization in amorphous Si up to temperatures close to T_{cl} (Sect. 3.2.3).

Apart from silicon, extensive work has also been done with other semi-conductor materials, in particular Ge and GaAs. Ge is found to behave very similarly to Si in most respects. The threshold for surface melting in amorphous Ge for ruby-laser pulses is less than 0.2 J/cm^2, and complete regrowth is observed in (100) material at 1.0 J/cm^2 [4.37]. The lower thresholds are explained by a somewhat larger absorption coefficient than Si, as well as by a lower melting point. The melting point of amorphous Ge was estimated to be 969 K [4.32], as compared to 1210 K for the crystal. Amorphization of Ge by laser irradiation has so far not been demonstrated.

The case of compound semiconductors like GaAs is somewhat more complicated, mainly due to their tendency to lose the more volatile component by preferential evaporation at elevated temperature. Furnace annealing is usually done by using surface encapsulation, usually with layers of SiO$_2$ or Si$_3$N$_4$, to prevent evaporation losses, but such layers complicate the coupling of optical beams and have often been omitted in laser remelting studies. GaAs forms a metallic melt [4.38]. Single-crystal regrowth is achieved with ruby-laser pulses of about 1 J/cm^2 [4.39], but the regrown material tends to contain dislocations and other defects. These disappear at higher fluences, but at the cost of increasing As loss, resulting in excess Ga ag-glomerates on the surface which must be removed by etching [4.40]. The concentration of electrically active dopants was found to be higher than in furnace-annealed material, but the carrier mobilities tend to be lower [4.41]. Attempts to activate low-dose implants (which do not amorphize the material) have met with only limited success. Similarly mixed results have been obtained with other III-V or II-IV compound semiconductors like GaP, InP, or Cdse. For a review, see [4.42].

We have so far mainly dealt with lattice regrowth in the ion-implanted amorphous semiconductors. Let us now consider what happens to the im-planted species during the rapid crystallization process.

4.2.2 Segregation and Trapping

It is well known that impurities in an undercooled melt tend to be segregated from the advancing crystal boundary – this is the basis of zone refining. The amount of segregation depends on the *interfacial distribution coeffi-cient* which gives the ratio of the impurity concentration at the crystal boundary to that in the melt adjacent to it

$$k = (X_s / X_l)_{interface} . \tag{4.17}$$

Here both X_s and X_l are assumed to be small (typically a few at.% or less) whence k is practically independent of composition. At near-equilibrium, i.e, for slow growth, the concentrations X_s and X_l are those given by the solidus and liquidus lines of the phase diagram; their ratio is referred to as the *equilibrium distribution coefficent* $\bar{k}$. Usually $\bar{k}$ is smaller than unity, indicating that the impurity is less soluble in the crystal than in the melt.

Consider now what happens to an implanted impurity distribution in a semiconductor irradiated by a short pulse. Figure 4.12 exhibits a sequence of numerically obtained distribution profiles [4.7]. The distribution of imp-lantation is roughly Gaussian in shape, with the position and width of the peak determined by the mass and energy of the implanted ions. As the semi-conductor is molten (t = 0), the distribution begins to spread rapidly by dif-fusion. However, as the interface (dashed vertical line) begins to return to the surface, only part of the impurity present there (10%, corresponding to k = 0.1 in the example) is absorbed by the crystal, while the remainder ac-cumulates in the melt ahead of the interface. The melt gets more and more enriched in the impurity, and when solidification is completed (t = 50ns) a considerable amount of it ends up close to the surface.

What this description ignores is the fact that in the end the melt is likely to be a concentrated mixture of two species. The assumption $X_l \ll 1$ will break down, and the distribution coefficient will not remain independ-

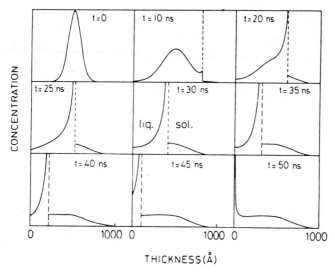

Fig.4.12. Numerically calculated impurity profiles in a sample with a Gaussian initial impurity distribution. The frames show instananeous profiles on both sides of the l-s interface (dashed line) during solidification. Parameters used: u = 2 m/s, D = 10^{-4} cm^2/s, k = 0.1 [4.7]

ent of the composition. Moreover, except for a few binary systems which form continuous solid solutions, there will always be a maximum value for X_s beyond which compound phases will form, rather than a solid solution. Also, interface instability and cellular growth are often observed near the surface. In the next section we shall deal with this and other solidification phenomena in concentrated mixtures. Let us now focus on that part of the final distribution in Fig.4.12 where X_s is still small.

Dopant redistribution during laser remelting and the resulting concentration levels of dopants in Si have extensively been investigated, and conventional models have not always been found adequate to explain the results. As an example, Fig.4.13 plots measured concentration profiles in In-implanted Si before (open circles) and after (closed circles) regrowth by a 15 ns ruby-laser pulse. Also shown are fitted profiles in which the distribution coefficient was treated as a fitting parameter. Some surface accumulation of In is evident, but, more importantly, the best-fit value of k turns out to be 0.15, three orders of magnitude above the equilibrium value! The concentration of substitutional In atoms is $5 \cdot 10^{19}$ cm^{-3}, exceeding the equilibrium solid solubility limit by a factor of 60. Similar results have been obtained for all Group-III and -V dopants, each element showing characteristic enhancement factors for k and the maximum solubility [4.43]. Growth direction also plays a role – k values for growth along ⟨111⟩ tend to be systematically larger than for growth along ⟨100⟩ [4.44]. Increased solubility under ns laser remelting was also observed for Pt [4.8], but not for

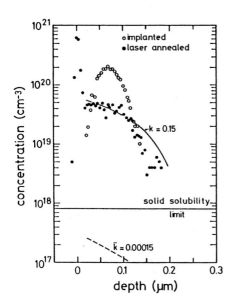

Fig.4.13. Rutherford backscattering spectra of In distribution in ion-implanted and laser annealed Si, showing supersaturated solid solution. The peak at the surface (closed dots) is indicative of segregation. The dashed line gives the profile calculated using the equilibrium segregation coefficient $\bar{k}$, the solid line is a fit to the measured profile [4.43]

other transition metals like Cr, Cu, Fe, W, Zn which are nonsubstitutional in Si [4.45].

The formation of supersaturated crystals from rapidly cooled melts is, in itself, not surprising and not in contradiction to local equilibrium thermodynamics. Driving an interface at a larger velocity increases its undercooling (Sect.4.1.4) which usually leads to a larger solubility. However, the case of Si is special in that its equilibrium solubility curve for most Group-III and -V dopants is of the retrograde type. In retrograde systems the solubility, rather than increasing with the undercooling, has a maximum beyond which it decreases with further undercoooling. Supersaturation beyond a retrograde maximum, as observed in the laser remelted Si, indicates a deviation from local equilibrium at the interface. The effect is, following *Baker* and *Cahn* [4.46], known as *solute trapping*.

How does solute trapping work? Segregation, as apparent from Fig. 4.12, requires the rejected solute atoms to move ahead of the advancing interface by diffusion. The idea of solute trapping is simply that the interface moves faster than the solute atoms, which are thus engulfed by the interface and get buried in the crystallizing solvent. The interface velocity u at which this should occur can be estimated from the melt diffusivity. The minimum time an impurity atom needs to diffuse out of a monolayer of thickness a ahead of the interface is a^2/D, while the crystal grows by the same amount in a time a/u. Impurity trapping should occur when $a^2/D \simeq a/u$, or $u \simeq D/a$. Setting, for an order-of-magnitude estimate, $a = 2 \cdot 10^{-8}$ cm and $D = 10^{-5}$ cm²/s yields a critical interface velocity of 5 m/s, in rough agreement with the experiment. Impurity trapping is, incidentally, also observed during solid-state regrowth of ion-implanted Si [4.47]. Here velocities are many orders of magnitude slower than in pulsed remelting, but the difference is made up by the smaller diffusivity in the amorphous phase.

Supersaturation destabilizes the crystal lattice due to the stress created by a large number of substitutional foreign atoms. For example, boron which has a smaller covalent radius than Si, causes the Si lattice to contract by an amount proportional to the local boron content, and the Si cracks for concentrations exceeding 4 at.% [4.48]. Sb, which has a larger radius than Si, causes a lattice expansion instead [4.49]. Thermodynamically, the destabilization of the lattice is expressed by an increase in the free energy of the crystal. There is a thermodynamic limit for supersaturation, given by the criterion that the free energies of the solid and the liquid become equal, i.e., that the driving force for crystallization vanishes. The corresponding concentrations are those at which the free-energy curves of the solid and the liquid cross (Fig.4.8). Plotted as a function of temperature on the phase diagram, these concentrations form a curve located somewhere between the liquidus and solidus lines, known as the T_0 curve. The T_0 curve indicates the theoretical maximum of supersaturation achievable by growth from the melt. Estimates indicate that the values observed for Group-III and -V dopants in laser annealed Si are indeed close to the theoretical limit [4.50].

It is quite obvious that the simple binary growth model mentioned in Sect.4.1.4 with independent component growth rates is unable to describe solute trapping, in which one species is buried in the other. A number of kinetic models of solute trapping have been put forward [4.50-53]. These models are based on specific assumptions about the growth process and the structure of the interface, amounting to various forms of relationship between the component growth rates. *Jackson* and coworkers essentially added a term proportional to $u \cdot X_1$ to the solute freezing rate [4.51], while *Wood* took the activation energy for solute remelting to increase with increasing interface velocity [4.52]. Figure 4.14 illustrates the variation of k with interface velocity for various dopants in Si, calculated from *Wood's* model [4.52]. *Aziz's* calculation allowed the solute to be trapped and to remelt at rates determined by the interface velocity and the diffusivity [4.53]. His segregation constant for a continuous growth process is of the form

$$k(u) = \frac{\beta + \bar{k}}{\beta + 1} \qquad (4.18)$$

where $\beta \equiv ua/D$. Note that k reaches unity only if the interface velocity u becomes large compared to the "diffusive" velocity D/a, as intuitively expected from the qualitative argument given earlier. The finding that for $\langle 111 \rangle$ growth in Si k is larger than for $\langle 100 \rangle$ growth is consistent with this picture, as a larger undercooling is required for a given interface velocity in the $\langle 111 \rangle$ case. Hence the melt diffusivity should be smaller, favoring trap-

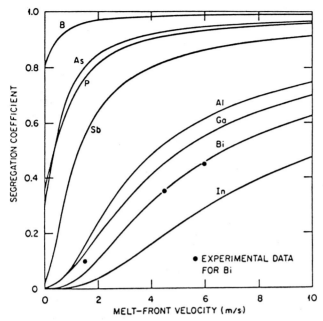

Fig.4.14. Calculated variation of the segregation coefficient with the l-s interface velocity. Some experimental data are shown for comparison [4.52]

ping. Segregation and trapping are even observed if the amorphous, rather that the crystalline phase grows form the melt, with similar values of k(u) [4.54].

4.2.3 Metallic Substrates

Stimulated by the intriguing results obtained with Si, many researchers have also applied ion-implantation followed by surface remelting (by laser or electron beams) to metal single-crystals, such as Al [4.55], Cu [4.56], or Ni [4.57]. The ultimate goal here is not doping but the improvement of corrosion or wear resistance of metallic surfaces. The results obtained so far, although not as complete as those for Si, tend to reveal qualitatively the same phenomena of epitaxial regrowth, impurity segregation and trapping as found in semiconductors. Important differences between metals and semiconductors exist with respect to the coupling parameters as well as the thermal data (Appendix A). Metals have high reflectances which tend to decrease with temperature, and choosing the correct fluence to melt without damaging them is often more critical than in Si. Further, the absorption lengths in metals are extremely short and lead to very large volumetric heating rates and large thermal gradients. Regrowth velocities for equal pulse durations are also appreciably larger than in Si because thermal conductivities are larger and latent heats are smaller. For example, calculated interface velocities in Si and Al irradiated by 50 ns electron-beam pulses (for which differences in the absorption length are insignificant) are 1.7 m/s in Si and 8.3 m/s in Al [4.58].

Regrowth after pulsed melting in pure as well as ion-implanted single crystals has been studied in some detail for the case of Al [4.55]. Electron rather than laser-beam pulses have often been used in this work, but the metallurgical results are not sensitive to the method of energy deposition. The most conspicuous difference to regrowth in Si is that regrown metal crystals contain numerous defects and never quite reach the quality of the virgin crystal, even in pure specimens (for this reason the term *laser annealing* is rather misleading in the case of metals). Metals have high equilibrium densities of vacancies at elevated temperature. Large densities of vacancies also appear to be introduced during solidification. Upon rapid cooling the vacancies are quenched-in but remain mobile enough at ambient temperature to coalesce into dislocation lines or loops in ways that depend on the implanted species [4.59]. In addition, thermal stresses present during and after resolidification can lead to slip deformations. As an illustration, Fig.4.15 shows slip bands in (110) Al after melting by a 20 ns laser pulse [4.60]. Slip occurs along the (111) planes, and large densities of dislocations are present between the slip traces.

Differences between metals and semiconductors also exist with respect to segregation and trapping. Extended solid solubility by trapping tends to be pronounced even more than in Si, mainly due to the larger regrowth velocities. *Picraux* et al. [4.58] have compared segregation of implanted Sb in Al and Si, which both show retrograde solubility for Sb, under identical conditions. Enhancements of the distribution coefficient were about a factor of 10 in Si, but a factor of 250 in Al. Residual segregation was observed in Si, but complete trapping (corresponding to k = 1) in Al. The difference was explained by the larger interface velocity in Al, as well as by the fact that the diffusity of Sb in liquid Al is only about one tenth of that in liquid Si. Strongly supersaturated solutions in Al have also been obtained with a number of other implanted species, including Cu, Cr, Ni, Mo, Sn and Zn.

Complete trapping means that the impurity profile present in the melt at the instant of solidification is frozen-in unaltered and thus available for quantitative analysis after solidification. Such analysis has yielded interesting insights into the transient diffusion process acting in the short-lived melt prior to solidification. The simple diffusion equation (4.1) is strictly valid only for diffusion under isothermal conditions, while in pulsed-melting diffusion is far from isothermal. Heat and mass flow occur simultaneously within the same volume, and they interact [4.61]. In a binary mixture kept in a constant thermal gradient one observes the spontaneous buildup of a

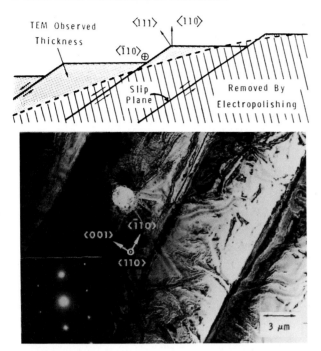

3.5 J/cm^2, 20 ns

Fig.4.15. Electron micrograph taken normal to the surface, and schematic cross section showing slip bands in (110) single-crystalline Al after laser melting [4.60]

compositional gradient (typically with the heavier component accumulated at the cooler end), such that at steady state the two gradients are related by

$$\frac{\nabla X}{\nabla T} = X(1 - X)S_T \ . \tag{4.19}$$

Since the total particle flux vanishes at steady state, there must be a contribution to the particle flux caused by the thermal gradient. This is known as the *thermodiffusion*, or *Soret effect*.[1] The Soret coefficient S_T is typically of the order of 10^{-5} to 10^{-3} K^{-1} and may depend on temperature [4.62]. Neglecting this dependence, the corrected diffusion equation can be written as

$$\frac{\partial X}{\partial t} = \nabla D[\nabla X + X(1-X)S_T \nabla T] \ . \tag{4.20}$$

---

[1] There is also a contribution to heat flow from the concentration gradient (Dufor effect), but it is not significant in liquids.

95

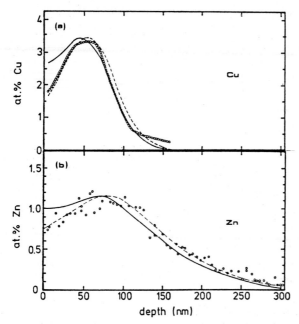

Fig. 4.16a,b. Measured (dots) and calculated (lines) concentration profiles of ion-implanted Cu and Zn in Al after laser irradiation. Dashed and solid lines indicate profiles calculated with and without allowance for thermodiffusion, respectively. The value of S_T used in (a) is $2.5 \cdot 10^{-2}$ K^{-1}, that in (b) is $1 \cdot 10^{-2}$ K^{-1} [4.63]

Thermodiffusion, while in principle a second-order effect, can be significant in laser remelting where thermal gradients are large. Figure 4.16 exhibits measured and calculated Cu and Zn profiles in ion-implanted and laser-irradiated Al [4.63]. Calculated profiles were obtained from (4.20) with and without allowance for thermodiffusion, with S_T treated as a fitting parameter (the Soret coefficients determined by *Dona Dalle Rose* and *Miotello* are net values, averaged over the whole thermal cycle [4.63]).

4.3 Surface Alloying

The basic sample structure in surface alloying is an overlayer on a substrate (Fig. 4.1b). Alloying is achieved by melting the overlayer as well as a comparable thickness of the substrate. This geometry allows a higher concentration of admixture at a depth within a substrate larger than is achievable by ion implantation. Rather than with "host" and "impurity" species, we deal here with mixtures of substrate and overlayer material in comparable concentrations. In this situation (as opposed to that of the previous section)

96

the type of phase diagram of the substrate-overlayer system is important in explaining the microstructure of the alloy. However, due to incomplete interdiffusion, the alloyed layers tend to be inhomogeneous even on a macroscopic scale (in contrast to those of the next section). This basic inhomogeneity further influences the microstructure of the alloys.

4.3.1 Semiconductor Substrates

Many transition-metal silicides are metallic and yield Ohmic contacts of low resistance and good thermal stability in semiconductor devices. Conventional techniques of silicide formation employ furnace annealing of thin vapor-deposited metal films on Si, in which stoichiometric compounds grow by solid-state reactions (Sect.3.2). The laser-remelting approach avoids exposing the entire wafer to the high temperature required for the solid-state reaction. Pulses in the ns regime have been used in most of this work. The results tend to be rather different from those of solid-state compound formation.

The elemental distribution resulting from pulsed surface alloying can, in a first approximation, be understood from simple diffusion considerations. As a typical example, Fig.4.17 shows Rutherford backscattering spectra of two Pd films, 200 and 48 nm thick, on a Si wafer after irradiation by equivalent laser pulses [4.64]. The profile obtained with the thick film, where some unreacted material is left at the surface, closely resembles a diffusion profile. The thinner film is completely consumed by the alloying process, and the resulting profile is steplike. This indicates that in the first case diffusion is incomplete and limited by the melt lifetime, whereas in the

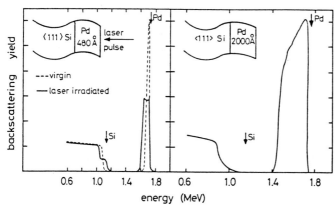

Fig.4.17. 2.0 MeV He$^+$ backscattering profiles of vacuum-deposited Pd films on Si(111). Dashed lines: As-deposited, solid lines: after laser alloying with a 1.6 J/cm^2, 18 ns Nd laser pulse (the arrows give the surface position of the respective atoms) [4.64]

second case it is complete and yields a final concentration determined by the melt depth. The threshold for the alloying process follows from the requirement that the melt depth exceeds the deposited film thickness. It depends on the thickness of the film as well as on the melting points of film and substrate material. For element combinations with a deep eutectic in the phase diagram, the threshold is lowered because melting sets in at the eutectic temperature.

Closer examination of the alloyed structures reveals a variety of complex microstructures not apparent from Fig.4.17. X-ray diffraction often shows not only one but several compound phases, often including metastable ones. For example, in the Pd-Si alloyed layers the compounds Pd_5Si, Pd_4Si, Pd_3Si, Pd_2Si, and Pdsi can be detected simultaneously, with concentrations depending somewhat on the original film thickness and on the laser fluence. The electrical resistivity of such mixtures can be a factor of 5 to 10 higher than that of a single-phase layer formed by solid-state reaction [4.65]. This does not necessarily render them useless as Ohmic contacts, however. Device-quality laser formed Ohmic and Schottky-barrier contacts have been demonstrated [4.66]. Satisfactory Ohmic contacts have also been obtained on n-type GaAs by laser alloying of deposited Ge layers [4.67].

The presence of several compound phases is not surprising in view of the fact that the local melt concentration prior to solidification ranges from pure Si to almost pure metal − covering, as it were, the whole phase diagram. However, the alloyed layers often also exhibit a complex lateral microstructure, consisting of "cells" of nearly pure Si surrounded by "walls" of silicide. The occurrence of cellular growth, which has been observed for a variety of metal layers on Si (including Pd, Pt, Ni, Fe, Co and Rh) is due to the effect of *Constitutional Supercooling* (CS).

4.3.2 Constitutional Supercooling

Under conditions of CS, an initially plane interface tends to develop into a corrugated array of columns or even more complex shapes [4.68]. The effect depends on a subtle interplay between macroscopic diffusion of heat and solute, and microscopic interface dynamics. To understand it, consider crystal growth from a diluted mixture in the presence of segregation. Segregation causes a buildup of impurity in the melt ahead of the liquid-solid (l-s) interface. Enriching a melt usually lowers its liquidus temperature and thus increases its undercooling, as indicated in Fig.4.18a. Under the conditions shown (liquidus temperature decreasing with increasing impurity content; gradient in liquidus temperature larger than gradient in actual temperature) the undercooling, and thus the "nominal" growth velocity, increase with distance away from the l-s interface. Crystal growth under such conditions is inherently instable: Random protrusion in the crystal boundary, due to fluc-

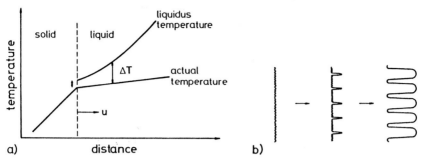

a)

b)

Fig.4.18. (a) Schematic temperature profiles in the presence of Constitutional Super-cooling. (b) Schematic sequence showing column growth from random fluctuations

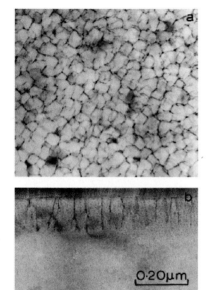

0·20μm

Fig.4.19. Electron micrographs showing (a) plain view and (b) cross section of cells in Sb implanted and laser annealed Si [4.71]

tuations, experience at their tip a larger undercooling and grow faster than their surroundings (Fig.4.18b). Eventually they develop into columns. The remaining melt is enriched in the impurity and trapped between adjacent columns. It eventually freezes out in the form of walls surrounding columns of relatively pure substrate material. An example of the resulting structure in the case of Sb implanted Si is shown in Fig.4.19.

A quantitative theory of CS developed by *Mullins* and *Sekerka* [4.69] predicts the critical impurity concentration for cell formation, as well as the cell size, to decrease as a function of interface velocity. The theory, which in its original form assumes local equilibrium at the interface, has been adapted by *Narayan* [4.70, 71] to allow for a velocity-dependent distribution coefficient. In this modified form the theory appears to give correct predic-

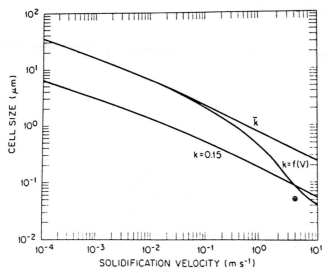

Fig.4.20. Calculated cell size at the onset of instability due to Constitutional Supercooling in Si containing In. The three curves are obtained by using for the distribution coefficient the equilibrium value $\bar{k}$, the measured value 0.15 (Fig.4.13), and a function of interface velocity, f(v), respectively. The dot shows an experimental point [4.70]

tions of the critical impurity content as well as the resulting cell size in laser-annealed implanted Si. As an illustration, Fig.4.20 shows calculated cell sizes as a function of interface velocity in In-implanted Si.

Constitutional supercooling depends on the composition gradient in a melt, but the gradient need not be due to segregation – it arises naturally from interdiffusion of two elements in laser alloying. It is fairly obvious that instability is likely to occur whenever the melting point of the substrate material, from which solidification starts, is higher than that of the overlayer or of any intermediate phase. The concept is sketched in Fig.4.21: The concentration profile, e.g., that shown in Fig.4.1b, is combined with the phase diagram to yield a depth profile of the liquidus temperature. The occurrence of CS can now roughly be predicted by adding transient temperature profiles to the plot: Whenever the undercooling increases away from the momentary position of the interface, growth instability is expected [4.72]. The scheme predicts CS in all cases where cells have been observed. No CS is predicted, e.g., for Mg on Si, where indeed none is observed [4.73].

The type of instability illustrated by Fig.4.21 is somewhat different from that due to segregation alone, and it may or may not result in patterns as simple as those of Fig.4.19. Since a large volume of melt tends to be simultaneously undercooled in the present case, nucleation and growth of compounds, or of the pure overlayer material, may occur in the melt ahead of the "main" interface, perhaps concurrently with cell development by the lat-

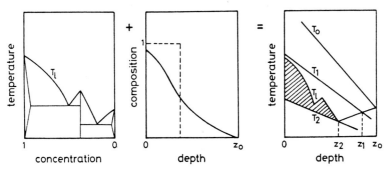

Fig.4.21. Schematic illustrating the occurrence of constitutional supercooling during regrowth of samples of the type of Fig.4.1b. T_0, T_1, and T_2 are subsequent temperature profiles in the melt. The hatched region is undercooled [4.72]

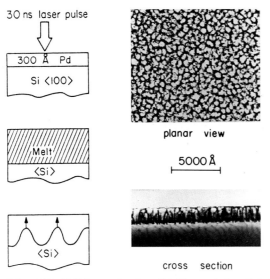

planar view

5000 Å

cross section

Fig.4.22. Cell formation in a Pd-Si sample irradiated by a Q-switch laser pulse [4.72]

ter. The result can be a complex or even apparently random mixture of phases. The case of Pd on Si is illustrated in Fig.4.22. Here the Si columns are epitaxial with the substrate, while the cell walls consist of a mixture of silicide phases. In the case of Pt on Si partially amorphous regions have been detected, along with a mixture of stable and metastable compound phases [4.65, 73]. Co and Ni, on the other hand, form single phase disilicides which, even though in a cellular arrangement, can be coherent with the Si matrix. Two different cell structures in Co and Si were described by *van Gurp* et al. [4.75]. Superimposed on a network of small (100 nm) cells due to CS were larger ($\simeq 1\,\mu$m) cells, which *van Gurp* et al. interpreted as Bénard cells, caused by a convective instability in the melt. As a source of

30 ns laser pulse

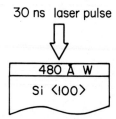

Fig.4.23. Frozen-in vapor bubbles in a W-Si sample irradiated by a Q-switch laser pulse [4.72]

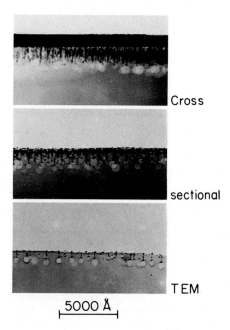

Cross

sectional

TEM

5000 Å

the instability they suggested a gradient in surface tension, arising from lateral variations in melt temperature caused by their sharply focused laser pulse. Yet another instability is observed in connection with refractory over-layers: They may crystallize while part of the melt below is still superheated [4.72]. Figure 4.23 illustrates the case of W on Si: Beneath a layer of poly-crystalline WSi_2 (etched away in the micrograph) there are myriads of tiny globular voids – frozen-in vapor bubbles an indication that the melting point of W is above the equilibrium boiling point of Si. Each bubble is connected to a stem of W-rich material, which apparently acted as a centre for heterogeneous nucleation of vapor in the superheated Si melt.

Interface instability is not limited to compound-forming systems but also observed for solid-solution formers and eutectic systems. Heteroepitaxy of Ge on Si(100) by laser alloying was demonstrated [4.76, 77], but cell formation and misfit dislocations prevented device quality heterojunction growth. The simple-eutectic systems Au-Si, Au-Ge, Ag-Si, and Ge-Al were investigated by *Lau* et al. [4.78]. A sample result for the first system is shown in Fig.4.24 (a thin amorphous Si layer on top of the Au layer served

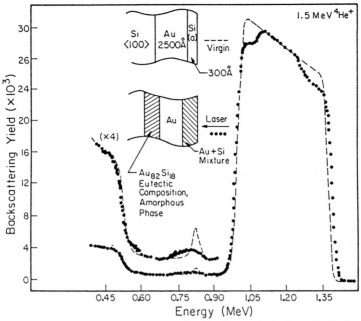

Fig.4.24. Backscattering spectra of a Au-Si sample before (dashed) and after (dots) laser alloying. The composition of the buried mixed layer is close to eutectic ($Au_{82}Si_{18}$) [4.78]

to facilitate absorption of the laser pulse). The main result is that the mixed layer between the Au film and the substrate, while varying in thickness with the laser fluence, always has a uniform composition close to the eutectic composition (18at.% Si), as long as not all of the Au is consumed. Analogous observations were made for the other systems. The interpretation is that a eutectic melt starts to form at the interface as soon as the eutectic temperature (370°C) is reached, i.e., while both the Au film and the Si substrate are still solid. Only at a much higher fluence does the elemental distribution start to resemble a diffusion tail, as in Fig.4.17. The structure of the eutectic alloy was amorphous at low fluence and crystalline at higher fluences (we come back to metastable Au-Si alloys in Sect. 4.4).

An obvious way to prevent growth instabilities is to eliminate the concentration gradient in the melt, e.g. by premixing or pre-reacting the sample, and to choose the composition within a one-phase region of the phase diagram. *Tung* et al. [4.79] used this approach to obtain epitaxial Ni and Co disilicides on Si. Homogeneous melts with compositions in a two-phase region, on the other hand, are known to solidify in a compositionally modulated pattern of periodic stripes called lamellae [4.80]. The stripes are formed by alternating regions consisting of the two phases compositionally adjacent to the melt composition. Submicrometer lamella formation has been demonstrated in CW Ar-laser melted Co-Si eutectic films [4.81, 82].

4.3.3 Metallic Substrates

The ultimate aim of surface alloying of metals is to improve the mechanical strength or the environmental stability of low-cost metals without having to invest large amounts of expensive transition metals as in bulk alloying. While the basic experimental approach is the same as in semiconductor alloying, the emphasis is quite different. Details of the microstructure are less important than the average composition of the surface alloy and its hardness, corrosion resistance and macroscopic integrity. The elemental distribution created by pulsed-laser alloying of metals is basically the same as in semiconducting substrates, and so are many of the microstructural features associated with segregation and CS.

However, while many experiments have used pulsed laser beams, the thrust for industrial applications is clearly towards using rapidly scanned high-power CW laser beams. With powers in the kW range and dwell times of tens of ms, the melt depths achieved can reach one mm. Mixing of the elements over such distances cannot rely on diffusion alone, but requires convection. The scanned beam creates lateral temperature gradients away from the beam axis as well as along the scan direction. The temperature gradients induce surface-tension gradients which drive convection currents, as sketched in Fig.4.25. An analysis by *Gladush* et al. [4.83] shows that the convection influences heat and mass flow in a 1 mm deep melt pool for absorbed laser powers exceeding as little as 10 W/cm^2. At a more typical 50 kW/cm^2 they estimate convection current velocities of several m/s, corresponding to Reynold's numbers much larger than 1. Such convection currents produce far more efficient mixing than possible by diffusion. Remnants of macroscopic convective cells are often still evident in the resolidified structure. The same convection currents are also responsible for the characteristic surface rippling of CW-laser remelted surfaces [4.83-84].

Current methods of overlayer deposition include vapor deposition, sputtering, plating (electro- or electroless), powder or paste coating, spraying and rolling-on of foils. An interesting alternative to predeposition of overlayers is to add the alloy metal during scanned-beam irradiation of the

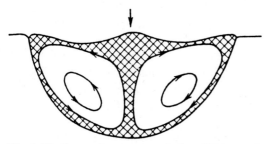

Fig.4.25. Convection currents during CW-laser alloying. The scan direction is perpendicular to the plane of the figure (schematic) [4.83]

substrate. This can be done by using a powder jet or a wire feeder nozzle close to the beam impingment area. For example, *Ayers* and *Tucker* [4.86] used this appraoch to inject TiC and WC particles into surface-molten steel. *Snow* et al. [4.87] described a technique termed "layer glazing", in which thick (up to one cm) alloys are built up layer-by-layer in repeated scans. The deposited material experiences a much larger cooling rate than achievable in casting, and its composition can be continuously varied by changing the feedstock. In a typical demonstration of this technique, *Snow* and co-workers deposited Ni-Al-Mo alloys on a circle along the circumference of a rotating cylinder, building up a disk about 15 cm in diameter, that was subsequently machined into a turbine wheel.

While the experimental techniques of laser surface alloying reached an impressive level of versatility, there are a number of physical constraints on what combinations of metals can be alloyed. The first is the type of phase diagram (Table A.5). The most unfavorable are Type 4 systems which form immiscible melts. Experiments with Ag-Ni [4.57, 87] have shown that even in such systems thin substitutional alloys can be achieved, but only if ns pulses are employed and if the overlayer (Ag in this case) is no thicker than some 10 nm. Thicker overlayers and longer irradiation times lead to phase separation. Generally speaking, successful alloying of Type 4 systems at large concentrations can be expected only if very high melt temperatures – above the liquid miscibility gap – can be sustained for times long enough to enable interdiffusion, followed by sufficiently rapid quenching to suppress melt separation. Here laser remelting offers potential advantages over conventional alloy forming processes.

Systems of Type 1-3 (Table A.5) do not pose particular obstacles to alloy formation from the thermodynamic point of view. Of particular interest are Type-2 and -3 systems in which the rapid cooling inherent in laser remelting enables the formation of small-grained – and therefore hard – alloys not available by other techniques. This is not the case for Type-1 systems which tend to form homogeneous substitutional alloys independently of the method of preparation.

Another physical contraint on the element combinations suitable for laser alloying is evaporation. Melting points and vapor pressures vary widely among the elements (Table A.6). Combinations of volatile and refractory metals necessarily lead to problems with evaporation losses or bubble formation. Serious bubble formation was observed, e.g., in laser-remelted cast Fe alloys containing phosphorous [4.89]. The amount of evaporation can apparently be influenced by formation of low-vapor pressure intermetallic phases. For example, strong Zr evaporation losses were observed in laser alloyed Zr layers on Al, Ti and V substrates, but less on Fe and not at all on Ni [4.90].

A significant number of alloy systems have been investigated by surface alloying to date and extensive bibliographical reviews have appeared in

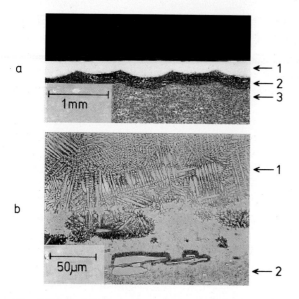

Fig.4.26a,b. Micro-section of a steel substrate, CO_2-laser remelted in sequential parallel scans at 1 m/s and 50 kW/cm². Zones: 1 - remelted, 2 - heat-affected, 3 - undisturbed. (**a**) Overview, (**b**) enlargement of the boundary between zones 1 and 2. Courtesy R. Dekumbis, EPFL Lausanne

the literature [4.91]. Particular attention has been devoted to Fe-based substrates, surface alloyed to elements such as Ni [4.85], Mo, V [4.92], Cr [4,92,93] and others. This work has mainly relied on scanned CW CO_2 lasers. A general result is that laser surface alloying of low-grade steel substrates can yield corrosion behavior equivalent to that of bulk stainless steel, but at a small fraction of the precious metal expenditure.

Detailed microstructural investigations have been performed on prealloyed steel substrates, laser remelted with or without additional overlayers [4.89]. Such investigations tend to reveal the presence of three well-defined microstructural zones beneath the laser track – (1) the remelted layer itself, (2) a region of heat-affected material, followed by (3) the undisturbed substrate material. A typical example is shown in Fig.4.26 for the case of a bare tool-steel. The sharp boundary between layers (1) and (2) marks the maximum penetration of the liquid-solid interface. It often coincides (in the case of a foreign overlayer) with a compositional discontinuity. Layer (2) consists of substrate material that merely underwent a solid-state thermal cycle. Its boundary with the substrate (3) represents the isothermal surface corresponding to the transformation temperature (Sect.3.2.5). Depending on the material composition and the local cooling rate, the material (2) can turn out to be hardened or softened with respect to the substrate. Both the remelted and the heat-affected layers exhibit characteristic microstructures.

In Fig.4.26, layer (1) consists of austenite (white), interwoven by a fine network of dendritic carbides (black), as is typical for rapidly solidified carbon steels. Zone (2) here contains mainly newly formed martensite, along with scraps of carbide (rodlike, near the lower edge of the figure) that have survived unaltered from the original material (3) – ferrite with interdispersed carbide.

Among the non-ferrous alloy systems that have been subjected to laser remelting there are examples from all the alloy types of Table A.5 [4.91]. Yet much remains to be done in this field, given the number of possible combinations of interest. Technological metals like Al, Cu, Ni or Ti have mainly served as substrates. Somewhat randomly selected examples of systems investigated are Cr and Pb on Al [4.95], Au, Sn and Ta on Ni, Ni-Cr on Cu [4.96] and Pd on Ti [4.97]. The high reflectance of many of these metals poses practical problems, particularly for infrared lasers and for partially overlapping scans. Highly heat conductive substrate materials, like Cu or Al, make alloying by CW beams more difficult than others, like Fe, Ni, or Ti, but produce higher cooling rates. Electron or repetitively pulsed laser beams may prove advantageous in such cases.

4.4 Melt Quenching

Laser remelting of structures of the type shown in Fig.4.1c yields a binary melt of homogeneous composition if the individual layers are thin enough. As an example, Fig.4.27 exhibits MeV He^+ backscattering spectra from a $Pt_{20}Si_{80}$ multilayer – prepared by sequential vapor deposition – on a sapphire substrate before and after remelting by a 50 ns laser pulse. Similar results are obtained with co-deposited or pre-alloyed layers, as well as by the use of homogeneous bulk alloys in place of a film/substrate configuration. The essential difference from alloying of an overlayer B with a substrate A is that now there are fixed supplies of A and B, such that the equilibrium configuration of the alloy is clearly defined by the phase diagram. However, the thrust of the experiments to be considered here is not towards equilibrium alloys, but towards glassy phases.

4.4.1 Glass Formation

Both nucleation and growth of crystals have significant rates only between the melting point and the glass temperature T_g. Cooling a melt through this range without nucleation and growth produces a glass, a frozen undercooled liquid. Substances like SiO_2 fail to crystallize if cooled faster than 1 K/h or so, while for metallic melts cooling rates between 10^6 and 10^{12} K/s or more

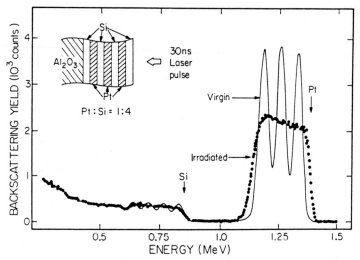

Fig.4.27. 1.5 MeV He$^+$ backscattering spectra of a Pt-Si film (80 at.% Si) on sapphire, before (solid line) and after (dots) laser irradiation [4.107]

are required for the same purpose. These enormous differences reflect different intrinsic nucleation and growth rates, arising for the reasons mentioned in Sect.4.1.

Metallic glass formation is performed today on an industrial scale almost exclusively by melt spinning. The available experimental work on laser melt quenching tends to fall into one of two categories: Fundamental investigations, mainly using short pulses, that explore glass formation in new materials, and more application-oriented work, typically employing rapidly scanned continuous beams and aiming at surface hardening of machine alloys. Cooling rates typical for scanned beams are comparable to those of melt spinning, and the results obtained with them have not been shown to be qualitatively different from those of mechanical quenching. We shall mainly emphasize the pulse work here.

Cooling rates in pulsed-laser quenching depend on the pulse duration and on the conductivity of the heat sink, but are quite insensitive to the type of material being quenched. Figures 4.28a and b exhibit numerically calculated cooling rates in films of $Au_{50}Ti_{50}$ alloy, deposited on sapphire or tungsten substrates, and irradiated by pulses of 50 ns and 50 ps FWHM duration, respectively (the film thickness must be adapted to the pulse duration for complete melting without evaporation). The cooling rates relevant for melt quenching are those below the melting point (dashed line) and are seen to be of the order of 10^9 K/s for the ns pulses and 10^{12} for the ps pulses. The use of refractory metal instead of sapphire substrates enhances the cooling rate by about a factor of three, due to a larger thermal conductivity. In Fig.4.29 phase diagrams of a number of systems quenched with 50 ns

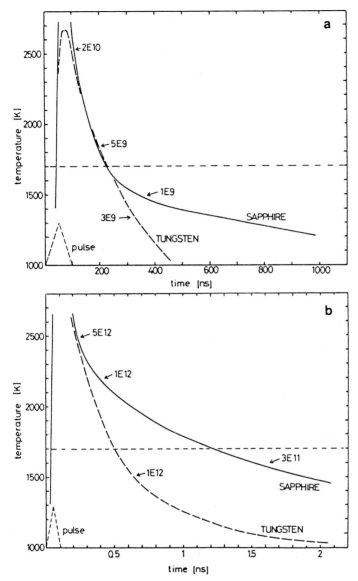

Fig.4.28. Numerically calculated surface temperatures in (**a**) 150 nm and (**b**) 60 nm thick $Au_{50}Ti_{50}$ films on tungsten and sapphire substrates, irradiated by (**a**) 50 ns and (**b**) 50 ps laser pulses. The dashed line indicates the melting point of the films. Exponential numbers indicate instantaneous cooling rates in K/s.

pulses are shown, with dark areas indicating the ranges where glass formation has been observed. The upper boundaries of the dark areas indicate the temperature at which crystallization of the glass upon slow heating was detected. Most of the glassy phases represented in the figure are not available at lower cooling rates.

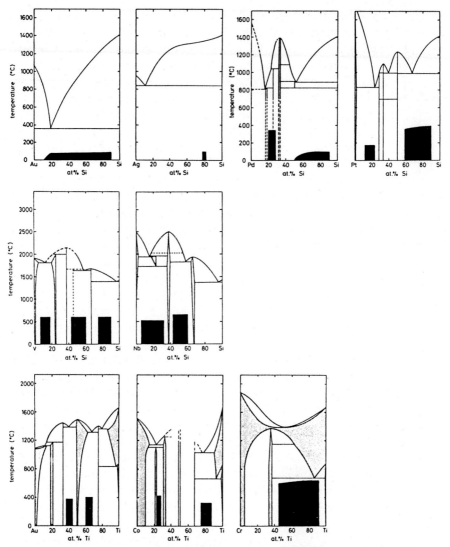

Fig. 4.29. Phase diagrams of some binary systems indicating experimentally established ranges of glass formation by ns-laser melt quenching of samples of the type of Fig. 4.27 (dark areas). Upper boundaries of dark areas give crystallization temperatures of glassy films upon slow heating. Dotted areas are regions of diffusionless crystallization.

Most of present knowledge about what is somewhat vaguely called *Glass Forming Ability* (GFA) has emerged from the experimental background of mechanical quenching, and hence pertains to cooling rates of the order of 10^6 K/s. Since short laser pulses produce cooling rates several orders of magnitude higher, it is not obvious that the same criteria should predict GFA by the two techniques. Kinetically, it is clear that the larger the reduced glass temperature $T_{gr} = T_g/T_{sl}$, the better the GFA. Good metallic-glass formers at 10^6 K/s have T_{gr} values around 0.5 to 0.6 [4.98]. A related criterion is the relative melting point depression of the mixture, $\Delta T_{mp} = (T_{av}-T_{sl})/T_{av}$, T_{av} being the weighted average of the component melting temperatures. ΔT_{mp} values of 0.2 or more indicate good GFA [4.99]. Obviously, both criteria favor compositions close to a deep eutectic. Neither is very successful in predicting GFA by laser quenching, as an inspection of Fig.4.29 shows. The same is true for several other empirical criteria of GFA considered in the literature [4.100].

Yet these are obvious regularities in the glass ranges shown in Fig.4.29 – they are all in two-phase regions of the respective phase diagrams, which are all of Type 2 or 3 (Table A.5). Continuous solid-solution formers do not yield glasses by ns quenching. An interesting limiting case is provided by systems like Cr-Ti (bottom right) which form solid solutions only at high temperature. Here the range of glass formation is very sensitive to the cooling rate – quenching with 50 ns pulses yields glasses between 45 and 65 at.% Ti for films on sapphire, but up to 85 at.% Ti for films on tungsten. These observations suggest that the requirement of long-range mass transport is the main hindrance to crystallization at cooling rates around 10^{10} K/s.

Diffusionless crystallization can become possible at large undercooling even in two-phase regions, however, due to formation of either supersaturated equilibrium phases or of "new" metastable compounds. Supersaturated solid solutions may form in systems with limited equilibrium solubility by means of trapping. The thermodynamic limits of trapping can be derived from free-energy considerations and quantified by means of T_0 curves (Sect.4.1.3). An example is the Ag-Cu system, which has a Type-3 (Table A.5) phase diagram but forms a continuous solid solution even at small undercooling [4.101]. Laser quenching with ns pulses in Ag-Cu produces the metastable solid solution instead of a glass. An example of a system forming a "new" metastable compound upon quenching is Ag-Si (Fig.4.24), in which glass formation is achieved only within a small window of composition near the Si-rich end of the phase diagram.

Complications with metastable crystals notwithstanding, it turns out that glass formation can be predicted quite reliably from standard thermodynamic data even at cooling rates around 10^{10} K/s. The idea is to construct Temperature-Time Transformation (TTT) plots for crystallization of every crystal phase competing with glass formation [4.102]. The crystallized volume fraction in a homogeneous melt cooled at t = 0 from T_{sl} to a temperature T below the nucleation temperature can be approximated by (assuming spherical crystallites)

$$\Xi(T,t) \simeq (\pi/3)Y(T)u(T)^3t^4 \quad \text{(for } \Xi \ll 1) \tag{4.21}$$

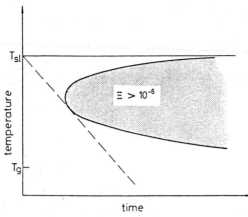

Fig.4.30. Schematic time-tempera-ture-transformation curve for for-mation of a 10^{-6} volume fraction of a crystal phase from an undercooled melt, according to equation (4.21). The dashed line indicates the critical cooling rate for glass formation.

where Y and u are expressions for the steady-state nucleation rate and the interface velocity, respectively, as given in Sects.4.1.4,5. Transient effects like the nucleation time-lag are neglected here. Glass formation is assumed to occur if Ξ remains below some detection limit, usually taken to be 10^{-6}. A schematic TTT curve indicating the region of detectable crystallization, as well as the limiting cooling rate to avoid it (dashed line), is shown in Fig. 4.30. The scheme can be extended to crystallization in a two-phase region by using the concentration-dependent driving forces introduced in Fig.4.8 [4.103]. The interface energy σ required to calculate the nucleation rate can be estimated from the latent heat of melting, while the jump frequency can be related to the melt viscosity [4.104]. To predict ranges of glass forma-tion, critical cooling rates are calculated for all possible crystal phases. "New" metastable compounds are treated analogously (this requires that suitable free-energy parameters can be estimated for them). Figure 4.31 dis-plays critical cooling rates calculated for the systems Au-Si and Ag-Si, tog-ether with the glass-forming range at the cooling rate used in the experi-ments [4.105]. The procedure correctly predicts the observed ranges of glass formation by laser quenching ($5 \cdot 10^9$ K/s) as well as those observed in mechanical quenching ($\simeq 10^6$ K/s). The basic physical reason for the differ-ence in GFA shown by the two systems is that the heat of mixing of liquid Au-Si is strongly negative, stabilizing the melt and hence the glass, while that of Ag-Si is positive at most compositions. It is clear that a strongly neg-ative heat of mixing of the melt is a generally favorable condition for glass formation, independently of the method of quenching.

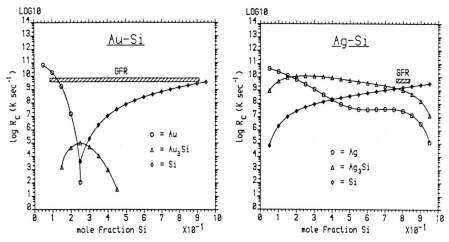

Fig.4.31. Calculated critical cooling rates as a function of composition for the systems Au-Si and Ag-Si, as determined by precipitation of the pure metal (o), the pure Si ($\diamond$), and a metastable silicide ($\triangle$). Also included are observed glass forming ranges at $5 \cdot 10^9$ K/s [4.105]

4.4.2 Silicon-Based Systems

Si and other Group-IV elements are known to facilitate glass formation in alloys. The first glassy metallic alloy ever to be produced by melt quenching was a eutectic mixture of Au and Si (18at.% Si), quenched by P. Duwez and co-workers in 1960 by means of a shock tube device at about 10^6 K/s.

The formation of pure amorphous Si by pulsed laser irradiation was discussed in Sect.4.2.1. Glassy Si-based alloys have been obtained by laser quenching with Q-switch pulses in a number of systems [4.105-107]. The crystallization temperature of the glassy films, as Fig.4.29 demonstrates, varies widely from system to system and even within systems. Au-rich glassy Au-Si films show precipitation of Au even at temperatures around $-30\,^\circ$C and must be kept cooled during and after irradiation, whereas Nb-Si glasses remain stable up to $600\,^\circ$C. Crystallization does not in all cases produce equilibrium phases. Metastable silicides form, e.g., in the Au-Si and Pt-Si systems from the glass, and the equilibrium configuration is reached only after annealing to a temperature $100\,^\circ$or $200\,^\circ$C higher than the crystallization temperature indicated in the figure.

Glassy Si-metal alloys offer a number of interesting features in their electronic properties. In metal-rich glasses of Au-Si, the Si atoms are found to be in a metallic state (pure solid metallic Si exists only at pressures above 120kbar). This is manifested in the electrical conductivity as well as in the reflectance spectra of the films [4.109]. Above about 30 at.% Si the metallic character of Si is lost, although the alloy still behaves as a true disordered metal. This holds for compositions up to about 85 at.% Si, where the material abruptly changes from metallic to semiconducting. The critical composition of 85 at.% is in accordance with Mott's criterion for a metal-nonmetal transition in this system. Altogether, the resistivity of the laser-quenched glassy Au-Si alloys can be varied via the composition over four orders of magnitude. Similar behavior is also found in other Si-metal systems.

4.4.3 Metal-Based Systems

By a metal-based system we mean a system in which nonmetals (including Si) are either absent, or present only as traces to facilitate glass formation, as often used in mechanical quenching. Glass formation in a number of metal-based systems has been achieved with scanned high-power CO_2-laser beams. Examples include alloys such as $Fe_{80}B_{16}Si_4$ [4.110], $Fe_{40}Ni_{40}P_{14}B_6$ [4.111], $(FeCr_{12})_{80}(C, B)_{20}$, Ni_xNb_{100-x} ($x = 30 \div 60$) [4.112], $Pd_{78}Cu_6Si_{16}$ and $Fe_{61}Cr_{10}Mo_5P_{16}C_8$ [4.113]. These alloys are "easy glass formers" well-known from mechanical quenching. The samples are usually either crystalline bulk alloys or powders applied to a steel base. Laser powers employed range from several hundred W to a few kW, and scanning rates are in the range of 0.1 to 1 m/s. With typical spot sizes of between 0.1 and 1 mm, dwell times are of the order of 1 ms. With these parameters, cooling rates are of the order of 10^5 to 10^6 K/s, i.e., the same as those typical in mechanical quenching. The thickness of material amorphized is given by the melt depth and is typically between a few tens and a few hundreds of μm. The technique has interesting potential for the surface hardening of machine parts [4.114] or the production of corrosion-resistant coatings, and it may also yield acceptable throughput for large-scale fabrication. Certain problems do exist, however, such as the tendency of amorphized zones to recrystallize during subsequent partially overlapping scans, the cracking of amorphous layers on substrates due to thermal stress [4.115], as well as the tendency of overlayers to lose their GFA due to admixture of substrate material.

Glass formation at compositions beyond conventional ranges of GFA, or in systems not previously considered to be glass formers, has been achieved by ns- or ps-pulse laser quenching in a number of cases. Among these are iron-rich Fe-B alloys [4.116], Ni-Nb, Mo-Ni, Mo-Co, and Nb-Co alloys [4.117], Au-Bi alloys [4.118] and even pure Ga, the only metal so far reported to be quenched to a glassy phase [4.119] (an amorphous phase observed in nominally pure Al after irradiation by 15 ns pulses turned out to be impurity-stabilized [4.120]). Particularly interesting kinds of metastable phases have been discovered in Au-Ti and Cr-Ti alloys that were deliberately laser quenched at cooling rates just insufficient for glass formation. The metastable crystalline phases thus obtained revert spontaneously to an amorphous structure upon subsequent furnace annealing [4.121].

5. Evaporation and Plasma Formation

The subtle and varied structural modifications considered in the previous chapters represent only a narrow window of irradiance or fluence, in which material is heated or melted but (essentially) not vaporized. In the following, material effects pertaining to the open-ended interval of irradiances exceeding the evaporation threshold will be considered. The physics involved is quite different from that considered in Chap. 4, and so is our point of view: Rather than studying what are essentially relaxation phenomena occurring after irradiation, we shall again deal with true light-matter interactions and consider relaxation effects only occasionally. The material effects in this regime are also rather less subtle than those discussed so far: Instead of microscopic structural rearrangements we shall be concerned with the transport of macroscopic masses over macroscopic distances, with ablation and compression of material and with internal energies large compared to chemical activation energies.

The main topic of this chapter, material evaporation, is the most traditional branch of laser processing. Related research activities started in the early sixties, soon after the first high-power ruby lasers became available. The early work was mainly motivated by the prospect of using laser beams as machining tools. Most of the phenomena governing evaporation by laser beams were understood by 1975, although, of course, important insights and refinements have been made in the meantime and continue to be made. Present research activities are mainly centered in two fields: improvements in the understanding and control of machining processes, and studies of the interaction mechanisms of ultra-intense radiation with matter.

Speaking of applications, machining – in all its variations from cutting through welding to milling and drilling – was the first area in which laser processing crossed the threshold to widespread industrial use. More prospective areas of application are shock treatments of materials and material testing under extreme conditions. Somewhat elusive remains the production of ultra-dense plasmas capable of sustaining thermonuclear fusion for energy production. Dense and hot laser-produced plasmas promise other uses, however, e.g., as intense sources of X-ray radiation.

This chapter is organized loosely following a sequence of increasing irradiance. We start again with a section discussing fundamentals. Section 5.2 then deals with "normal" material evaporation in which the laser beam just happens to furnish the energy, along with its main applications in machining. Section 5.3 considers irradiances sufficient to cause substantial ionization of the vapor and the surrounding atmosphere, while Sect.5.4 covers the highest available irradiances at which the very distinction between the condensed material and its vapor disappears. In the final Section 5.5 we return to solid earth, as it were, and discuss what has now become the most promising application of laser-produced vapors: their use as deposition sources for the production of thin films.

5.1 Fundamentals

5.1.1 The Thermodynamics and Kinetics of Evaporation

Since evaporation usually occurs from a liquid, let us start by considering the conditions for a liquid-vapor equilibrium. Phase equilibrium between a melt and its vapor requires equality of the free energies, but vapors are easily compressible and the equilibrium conditions now depend on the pressure. The change of free energy with temperature can be written as

$$\frac{dG}{dT} = \frac{\partial G}{\partial T} + \frac{\partial G}{\partial p}\frac{dp}{dT} \equiv -S + V\frac{dp}{dT} \tag{5.1}$$

where we have used the thermodynamic relation $dG = Vdp - SdT$. Eq.(5.1) must hold for every phase individually. At equilibrium the temperature and the pressure inside the liquid and the vapor must be the same, from which follows that

$$d\bar{p}/d\bar{T} = \frac{S_v - S_l}{V_v - V_l} \simeq \frac{\Delta H_{lv}}{T_{lv}\,\Delta V_{lv}} \tag{5.2}$$

where $\Delta V_{lv} = (V_v - V_l)$ is the volume change upon evaporation and where we have replaced the entropy change $\Delta S_{lv} = S_v - S_l$ by $\Delta H_{lv}/T_{lv}$, the ratio of the latent heat of evaporation to the liquid-vapor equilibrium temperature. Eq.(5.2) is the Clausius-Clapeyron equation which describes the dependence of vapor pressure on temperature, or, if written in the form

$$dT_{lv}/d\bar{p} = T_{lv}\frac{\Delta V_{lv}}{\Delta H_{lv}} \tag{5.3}$$

also specifies the dependence of the equilibrium temperature on pressure.[1]

---

[1] The values of the quantities T_{lv}, ΔH_{lv} and ΔV_{ls} will henceforth be understood to be those at 1 atmosphere, unless noted otherwise.

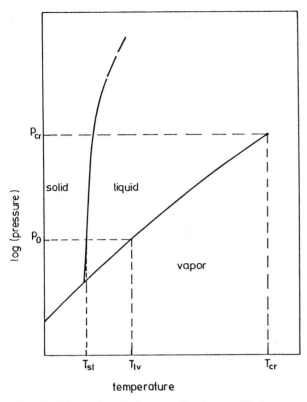

Fig.5.1. Schematic p-T diagram showing equilibrium ranges of existence of the solid, liquid and vapor phases

The transition between a liquid and its vapor at ordinary temperature is a first-order phase transition, like that between a crystal and its melt. The latent heat of evaporation is typically of the order of $2 \div 4$ eV per atom at normal pressure, as compared to 0.1 to 0.5 eV per atom for melting (Table A.4).

The equilibrium ranges of existence of the solid, liquid and vapor phases of an elemental substance are illustrated by the schematic pressure diagram in Fig.5.1. The liquid-vapor equilibrium curve, unlike the one separating the solid and liquid states, ends in a critical point beyond which there is no thermodynamic distinction between the liquid and the vapor. The critical temperature T_{cr} for metals is typically of the order of 3 to 4 times T_{lv}. At the critical point the latent heat as well as the volume change go to zero like $(T_{cr}-T)^{1/2}$, and hence the densities of liquid and vapor merge smoothly. Several material properties behave anomalously at the critical point; this is related to the fact that here $(\partial p/\partial V)_T$ vanishes. For example, density fluctuations become very large and the electrical conductivity drops sharply. Beyond T_{cr} "evaporation" proceeds as a continuous decrease in

117

density, without a phase transition and without thermodynamical instability.

An interesting problem exists with respect to the evaporation of metals, which involves the transition from a conducting to an insulating state. The transition is the inverse of Mott's transition and occurs at some transition density ρ_{mi} intermediate between the normal liquid and vapor densities. During subcritical evaporation the density changes discontinuously at the liquid-vapor boundary and "jumps" over ρ_{mi}, hence the liquid-vapor and the metal-insulator transitions occur simultaneously. But what, if the material is heated beyond the critical point and the liquid density decreases continuously to the vapor level? It is often assumed that ρ_{mi} coincides with the critical density ρ_{cr} [5.1], although there is no reason a-priori that this must be so. It has been hypothesized [5.2], that for certain metals $\rho_{mi} > \rho_{cr}$, i.e., the metal-insulator transition should already occur in the condensed liquid below T_{cr}. This would produce a transparent surface layer and strongly influence light absorption in the evaporating liquid. Experimental evidence for such an effect is, however, rather indirect and pertains only to the case of Hg [5.3]. Moreover, ionization of the vapor, which tends to obscure absorption phenomena in the dense phase, is possible at vapor pressures as low as a few atmospheres, far away from the critical regime.

Let us now return to normal evaporation. An expression for the equilibrium vapor pressure as a function of temperature is obtained by integration of (5.2)

$$\bar{p}(T) = p_0 \exp\left[\Delta H_{lv} \frac{T - T_{lv}}{RTT_{lv}}\right] = p_0 \exp\left[-\frac{\Delta G_{lv}}{RT}\right] \tag{5.4}$$

where p_0 is the ambient pressure. Eq.(5.4) is an approximation valid in the range $T_{lv} < T < T_{cr}$ of interest for the present discussion. It is based on the empirical fact that the ratio of the latent heat to the difference in compressibility between the vapor and the liquid is approximately constant in the mentioned temperature range [5.2]. An equation of the form (5.4), but with different parameters, is also valid at temperatures below T_{lv} (some vapor pressure data are given in Table A.6)

In nonelemental systems the liquid-vapor equilibrium obeys the same rules as that between a liquid and a crystal. Nonelemental vapors can always be regarded as ideal solutions, with the partial pressures given by $p^j = pX^j$ where X^j is the concentration of species j in the vapor. The partial pressures satisfy

$$\partial \bar{p}^j / \partial \bar{T} = \Delta h_{lv}^j / T\Delta V_{lv}^j \tag{5.5}$$

which generalizes (5.2). Here Δh_{lv}^j is the "partial heat of evaporation" of species j, (the enthalpy change per mole of solution contributed by evaporation of that species alone), and ΔV_v^j is the corresponding change in the partial molar volume. If the liquid is an ideal solution as well, then both quantities vary linearly with the melt composition.

As far as kinetics is concerned, there are two distinct modes of evaporation from a melt under laser irradiation: volume evaporation and surface evaporation. Surface evaporation is the normal case for metals, while volume evaporation may play a role in weakly absorbing media. Let us briefly consider the main aspects of the two modes of evaporation. The rate of surface evaporation can be obtained from considerations similar to those pertaining to the l-s interface velocity, but since we are mainly interested in the condensed phase (or what remains of it) we shall take a simplified approach here. Under equilibrium conditions the rate of particles evaporating from an open liquid surface is equal to the rate of particles condensing on it from their saturated vapor. The rate of condensation, and hence also evaporation, is obtained by integration over the Maxwellian velocity distribution of the vapor particles [5.4]. This yields a molar evaporation flux [moles/m² s] of

$$\bar{j}_v = \frac{A\bar{p}(T)}{\sqrt{2\pi MRT}} = \frac{Ap_0}{\sqrt{2\pi MRT}}\exp\left(-\frac{\Delta G_{lv}}{RT}\right). \tag{5.6}$$

Here A has the meaning of a "sticking coefficient", giving the fraction of vapor particles hitting the surface that adhere to it. For metals A is generally close to unity. If the actual pressure p in a liquid at temperature T is less than $\bar{p}(T)$, the liquid is superheated. However, the evaporation rate of a liquid should not depend on whether or not it is surrounded by its equilibrium vapor, and (5.6) is also valid for evaporation into vacuum or into an ambient atmosphere, provided p is not too large. The l-v interface velocity is then

$$u = \bar{j}_v V_d = M\bar{j}_v/\rho_d \tag{5.7}$$

where V_d and ρ_d are the molar volume, and the density of the melt, respectively. If a substantial vapor pressure develops, the interface velocity is reduced due to backflow and condensation of particles, as discussed in Sect. 5.2.

The superheated liquid also tends to boil. Boiling, or volume evaporation, requires nucleation and growth of vapor bubbles. The volumetric rate of steady-state homogeneous vapor nucleation can be put in a form equivalent to (4.15)

$$Y(T) = ANv_j e^{-\Delta g_N/kT} \tag{5.8}$$

where the jump frequency can be considered as purely impingement-limited, as in (4.11). Δg_N is the free energy of forming a critical vapor nucleus, obtained from (4.14) by replacing ΔG_{sl} and σ by the corresponding quantities for the liquid-vapor transition. In rapidly heated melts, vapor nucleation starts only after a time-lag similar to that observed in crystal nucleation [5.5]. Typical magnitudes for molten metals at moderate superheating are critical bubble radii of the order of $10 \div 100$ nm and time-lags of the order of 10^{-8} s. The strong temperature dependence of (5.8) again results in a sharp increase of the homogeneous nucleation rate above a relatively well-defined nucleation temperature T_N. However, boiling may develop at a much lower temperature as a result of heterogeneous nucleation in the presence of foreign particles (as Fig.4.23 nicely demonstrates).

Since a large change in density is associated with vapor nucleation, the strong increase in the nucleation rate above T_N causes violent expansion of the superheated liquid, called a *vapor explosion*. The energy liberated in a vapor explosion (the work done by the vapor) may reach a significant fraction of ΔH_{lv} (it may, in fact, exceed the specific energy of common chemical explosives [5.6]). In order for the liquid to reach T_N heating must be fast enough to prevent boiling by heterogeneous nucleation at small superheating. *Martyniuk* [5.6] estimated that heating rates of the order of 10^9 K/s are required to keep heterogeneous nucleation in liquid metals insignificant. Such heating rates are, of course, easily achieved with pulsed lasers in absorbing materials.

The composition of a binary vapor is, in general, different from that of the liquid in equilibrium with it, in accordance with the vapor-liquid phase diagram. In open evaporation from a surface the vapor is continuously removed, and only the freshly formed vapor is in contact with the melt. It is obvious that in this situation the melt is enriched in the less volatile species until either the more volatile component is lost or a congruently evaporating (or "azeotropic") liquid mixture is reached. However, this process depends on the presence of diffusional equilibrium in both the liquid and the vapor. If evaporation is fast, only a surface layer of the liquid is depleted in the volatile species, and the rate of evaporation of the latter drops quickly. At the same time backflow from the vapor (to be discussed in the next section) will mostly consist of atoms of the volatile species. The situation is reminiscent of the one discussed in connection with segregation in Sect.4.2.2 – in the absence of a diffusional equilibrium the vapor composition approaches the composition of the liquid. If backflow is neglected, this will happen after a time $\simeq D/ku^2$, D being the diffusivity of the volatile component in the melt, and k the distribution coefficient at the vapor-liquid interface. Remembering our discussion of trapping (Sect.4.2.2), we may again expect that k will tend towards unity if u approaches the "diffusive" velocity D/a, since particles of the less volatile species will be dragged along by the more volatile ones as the evaporating boundary overtakes diffusion in the melt. In a laser-irradiated solid where the melt is continuously formed ahead of the evaporating surface, the melt composition will itself be equal to the composition of the solid. Moreover, under powerful irradiation the atoms can acquire thermal energies far in excess of the binding energies, and differences in the partial enthalpies of the chemical species become unimportant. Hence strongly superheated solids will evaporate without a change in composition, and even the very presence of a liquid phase can usually be neglected. This feature is crucial in laser pulse-induced material deposition (Sect. 5.5).

We mention in passing that besides thermal (heat-driven) evaporation there are also nonthermal evaporation phenomena. Ultraviolet laser beams can induce optical excitation of molecules in some insulators, notably organic polymers, at rates rivaling or exceeding thermalisation. Since excitation tends to lower binding energies, excited particles may "evaporate", or desorb, before the material temperature rises appreciably [5.7]. This process, known as "photolytic desorption", explains why certain materials can be UV-ablated "cold" and at rates exceeding those expected for normal thermal evaporation [5.8].

5.1.2 Hydrodynamics

Hydrodynamics provides a point of view totally different from that of kinetics, useful particularly at intensities well above the evaporation threshold. The concept is to simply ignore the structure of both the material and also the liquid-vapor interface. The former is treated as a structureless fluid, characterized only by its density, pressure and energy content, and the latter as a mere discontinuity in the thermodynamic variables. We shall, as before, use the enthalpy to characterize the material in each phase. From the thermodynamic identity $H = U + pV$, where $U = \int c_v \, dT$ is the internal energy, the enthalpy $H = \int c_p \, dT$ can be written as

$$H = pV \frac{c_p}{c_p - c_v} \equiv pV \frac{\gamma}{\gamma - 1} \tag{5.9}$$

where $\gamma \equiv c_p / c_v$ is the adiabtic index. This procedure neglects the difference in temperature dependence of the specific heats, and for real substances averaged values for γ must be used. For a monatomic ideal gas $\gamma = 5/3$, and for a diatomic one $\gamma = 7/5$.

The familiar conservation laws of mechanics are now applied to relate the corresponding quantities on both sides of the boundary. Denoting with a subscript d quantities of the dense phase (no distinction is made between solid and liquid) and with v those of the vapor, we require that the flow of mass, momentum and energy across the boundary be continuous. For one-dimensional gas flow and neglecting viscosity we have, respectively, [5.9]

$$\rho_d v_d = \rho_v v_v , \tag{5.10}$$

$$p_d + \rho_d v_d^2 = p_v + \rho_v v_v^2 , \tag{5.11}$$

$$H_d/M + v_d^2/2 = H_v/M + v_v^2/2 . \tag{5.12}$$

Here the velocities[2] are measured with respect to the boundary (the velocity of the boundary in laboratory coordinates is $u = -v_d$), and we have used the mass density $\rho = M/V$ instead of the molar volume. If, e.g., ρ_d and H_d are known in advance, then the system (5.10-12) contains 6 unknowns. Using (5.9) for H_v, we need 2 more equations which have to be chosen according to the problem under consideration. Often the vapor is treated as an ideal gas for which the ideal gas law

$$pV = pM/\rho = RT \tag{5.13}$$

provides a connection between pressure, density and temperature.

A most important aspect of evaporation by a laser beam is the interaction of the evolving hot vapor with the beam itself. This interaction leads to strong nonlinearities in the coupling of an intense beam with a target, in addition to those that may arise in the condensed material.

5.1.3 Ionization of the Vapor

The equilibrium degree of ionization of a gas of density N and temperature T follows from the Saha equation

$$\bar{N}_e^2 = 2N \frac{g_1}{g_2} \left(\frac{m_e kT}{2\pi\hbar^2} \right)^{3/2} e^{-E_I/kT} \tag{5.14}$$

where g_1 and g_0 are the statistical weights of the ionized and neutral states, respectively, and E_I is the ionization potential. In a partially ionized gas, light is absorbed by thermally excited atoms (bound-free absorption), as well as by ions (bremsstrahlung absorption). The total absorption coefficient for $\hbar\omega \ll E_I$ can be estimated from the Kramers-Unsöld formula which holds for hydrogen-like gases [5.10]

$$\alpha = \frac{NZ^2 e^6 kT}{6\sqrt{3}\,\hbar^4 c\epsilon_0^3 \omega^3} e^{-(E_I - \hbar\omega)/kT} . \tag{5.15}$$

---

[2] In this chapter we use the symbol v for particle velocities and u for interface velocities.

More accurate calculations of absorption in hot vapors have been performed, e.g., for 10.6 μm radiation in Al vapor [5.11] and in metal oxide vapors [5.12].

With increasing irradiance the temperature and the enthalpy of the vapor increase. Light absorption further heats the vapor, which leads to even more absorption. This positive feedback favors the creation of a plasma in front of an evaporating target even at light fluxes far below the threshold for breakdown in a cold gas (see below). Typical irradiances at which this phenomenon is observed are 10^8 W/cm^2 for Nd lasers and 10^7 W/cm^2 for CO$_2$ lasers [5.13]. Associated vapor pressures are of the order of tens of atmospheres, corresponding to only moderate superheating of the liquid.

Once the gas is fully ionized, light absorption is dominated by bremsstrahlung absorption. For hot plasmas (kT $\gg$ $\hbar\omega$) the absorption coefficient can be written as [5.14]

$$\alpha(\omega) = \frac{N_i N_e Z^2 e^6 \ln(2.25 kT/\hbar\omega)}{6\pi\epsilon_0^3 m_e n_1 c\omega^2 kT\sqrt{2\pi m_e kT}} . \qquad (5.16)$$

This absorption coefficient varies like λ^2, and Fig.5.2 displays its dependence on temperature in fully ionized hydrogen plasmas at various densities [5.15]. Above the cutoff density ($\simeq 10^{21}$ cm^{-3}) where the plasma frequency exceeds the laser frequency, n_1 becomes very small (Sect.2.1), and the plasma behaves essentially like a metal. Below the cutoff density the absorption coefficient decreases strongly with increasing temperature, a feature that has important implications for laser-produced plasmas, as we shall discuss in Sect.5.4.

The hot vapor represents an easily ionizable medium, due to the thermal excitation of the atoms. However, even cold gases become ionized and absorbing in the presence of high irradiances, due to optical breakdown.

5.1.4 Gas Breakdown

Laser-induced gas breakdown involves the same mechanisms as breakdown in solid dielectrics. As discussed in Sect.2.2.4, ionization for $\hbar\omega < E_I$ can occur by multiphoton absorption or by avalanche (impact) ionization. Breakdown thresholds are usually determined by avalanche ionization, except for ultrashort pulses, or at gas pressures so low that the mean-free path between electron-atom collisions exceeds the focal diameter. Avalanche breakdown requires the presence of some "priming" free electrons, as in solid dielectrics. For the average energy gain of a free electron (2.44) still holds, but inelastic energy losses are negligible in the gas ($\delta E/E$ is of the order of the ratio of the electron mass to the atom mass). Furthermore, τ_e is

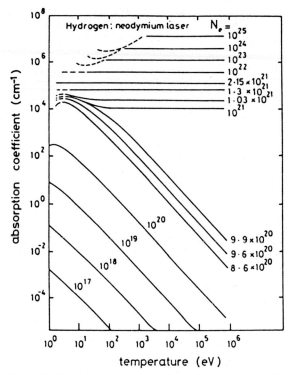

Fig.5.2. Calculated bremsstrahlung absorption coefficient as a function of temperature in a fully ionized neutral hydrogen plasma, for $\lambda = 1.06\ \mu$m and various electron densities. [5.15]

now the electron-atom collision time, which is inversely proportional to the gas pressure p. At normal pressures and optical frequencies we have $\omega \gg 1/\tau_e$ and, from (2.44), the breakdown threshold is expected to be proportional to τ_e and hence to $1/p$. A minimum in the threshold should occur at $\omega = 1/\tau_e$, which turns out to be at some hundred atmospheres for optical frequencies. Both predictions are borne out by the experiment [5.16]. For sharply focussed beams the breakdown threshold increases due to diffusion of electrons out of the focal volume, and an inverse relationship between the breakdown threshold and the focal volume results [5.17]. A detailed study of avalanche breakdown in air was presented by *Kroll* and *Watson* [5.18].

A key difference between avalanche breakdown in solid dielectrics and in cold gases is that priming electrons are very much scarcer in the latter. In normal air at STP the equilibrium density of ions is only about 10^3 cm^{-3}, and the natural ionization rate (by cosmic radiation, etc.) is no more than some 10 cm^{-3}s^{-1}. The probability of finding a free electron within the volume of a focal spot is thus negligible. The first electrons of the avalanche have to be delivered either by absorbing impurities of low-ionization potential, like dust particles, or by multiphoton ionization of gas atoms. In particular, gas-breakdown thresholds are found to be dramatically lowered by the presence of an absorbing target in the beam path. Here adsorbed impurities, which are evaporated and ionized far below the melting point of the target material, are almost always present. In practice, air breakdown thresholds in the vicinity of targets are of the order of 10^7 W/cm^2 for CO_2 lasers and 10^9 W/cm^2 for Nd lasers, roughly two orders of magnitude lower than the corresponding thresholds in clean air.

5.2 Evaporation at Moderate Irradiance Levels

The subject of this section is the regime of normal evaporation, in which the laser beam merely heats the condensed phase. This means, in particular, that the vapor is cool enough to be essentially transparent. Important laser applications in machining are often performed in this regime.

5.2.1 Beam Heating and Evaporation

A lower bound to irradiances of interest here may be estimated from the energy flux required for the surface to reach T_{lv} while the irradiation lasts, namely

$$It_p (1 - R)/d > [\Delta H(T_{lv}) + \Delta H_{lv}]/V_d . \tag{5.17}$$

Here $\Delta H(T)$ is the enthalpy difference of the material between ambient temperature and temperature T, and d is the larger of the absorption length α^{-1} and the thermal diffusion length $2(\kappa t_p)^{1/2}$. For a typical metal, taking 50 kJ/cm^3 for the RHS of (5.17) gives, e.g., for

t_p	1 s	1 ms	1 μs	1 ns
I(1–R)	>32 kW/cm^2	>1 MW/cm^2	>32 MW/cm^2	>1 GW/cm^2

The incident irradiance required in the case of metals would be a factor of 10 or so higher because of reflection. Since for the shorter pulses these irradiances approach or exceed those causing vapor ionization or even gas breakdown, the regime of normal evaporation is mainly relevant for pulse durations or dwell times well exceeding 1 μs. Nevertheless, most of the discussion which follows still applies even in the presence of vapor ionization, although the connection between incident irradiance and absorbed heat flux then becomes more complicated.

While it is understood that evaporation usually occurs from a liquid, we shall neglect the solid-liquid transition in the following. This is justified since the latent heat of melting is but a small fraction of the total heat content of a material at the evaporation temperature, as apparent from the enthalpy values listed in Table A.4. The heat-flow problem, as far as the dense phase is concerned, is thus reduced to that of Sect. 3.1, but there is an important new aspect: evaporation cools the surface. Every particle escaping from the dense phase carries along its binding energy (given by $\Delta H_{lv}/N_A$) plus some excess energy. On the other hand, as we shall discuss later on, there is also some backflow of particles from the vapor by which the cooling effect is reduced. If we neglect the excess energy as well as the backflow for the moment, the heat flux carried away by the vapor is given by $j_v \Delta H_{lv}$. A simple way to allow for this heat flux is to regard it as due to a "surface heat sink", to be treated in the same way as a heat source. Similarly, boiling could be described in terms of a "volume sink". However, volume evaporation causes strong expansion, not readily treatable by the formalism of Sect. 3.1. We shall limit this discussion to surface evaporation.

It is convenient to distinguish, as in Sect. 3.1, between surface heating (surface source) and volume heating (penetrating source). The first approach was shown to be adequate whenever the absorption length α^{-1} is small compared to the width $2(\kappa t)^{1/2}$ of the heated layer. This generally holds in metals as well as in nonmetals after breakdown. If heating is described in terms of a surface source, cooling by surface evaporation is allowed for by taking the net source flux as the difference

$$\Phi = I_a - j_v \Delta H_{lv} \ . \tag{5.18}$$

During the initial stages of beam heating, the temperature profile is described by (3.8). Once the surface approaches T_{lv}, material begins to evaporate and the absorbing surface – and hence the surface source – starts to move into the material with a velocity given by (5.7). For constant irradiance a steady state with a constant surface velocity will eventually be established. The temperature distribution inside the condensed material is then given by (3.21), with I_a replaced by Φ. The regime of steady-state evaporation by an extended source is adiabatic in the sense that the heat conducted away from the surface is not lost but eventually recovered by the advancing interface. In this regime the volumetric energy content of the blown-off material per unit time equals the absorbed irradiance, i.e., the steady-state interface velocity must satisfy

$$u = I_a \frac{V_d}{\Delta H(T_d) + \Delta H_{lv}} \tag{5.19}$$

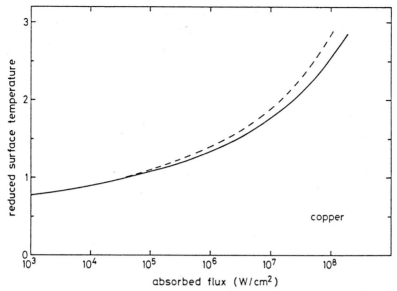

Fig.5.3. Reduced surface temperature T/T_{lv} during adiabatic evaporation as a function of absorbed irradiance for a Cu target. The dashed line is calculated from (5.20); the solid line also allows for backflow and excess energy of the vapor, according to (5.24) and (5.25)

where T_d is the surface temperature. The surface temperature in the adiabatic regime can be calculated by equating (5.7) to (5.19), which gives

$$\exp\left(-\frac{\Delta H_{lv}}{RT_d}\right)\frac{\Delta H(T_d) + \Delta H_{lv}}{\sqrt{T_d}} = \frac{I_a\sqrt{2\pi MR}}{p_0}\exp\left(-\frac{\Delta S_{lv}}{R}\right). \tag{5.20}$$

This shows that, as a result of the heat loss through evaporation, the dependence of temperature on irradiance is essentially logarithmic, rather than linear. The dashed line in Fig.5.3 gives the surface temperature as a function of I_a calculated from (5.20), using parameter values for copper (the solid line corresponds to a refined model discussed below).

The time required for establishing the steady state may be estimated from the time it takes the distribution (3.8) to reach the width κ/u of (3.21). This gives a build-up time for constant I_a of approximately κ/u^2. In the typical metal considered at the beginning of this section, u would be 2 mm/s for $I_a = 10$ kW/cm², and 20 cm/s for 1 MW/cm². The steady state would be reached after about 1 ms in the latter case, but only after as long as 10 s in the former. Evaporation at small irradiances is thus, in practice, always far from adiabatic, and most of the laser energy is spent in heating the solid, rather than in evaporating it. Adiabatic evaporation also requires that the heat flow is one-dimensional. This can be considered to hold if the width

κ/u of the heated layer is small compared to the beam diameter. Thus evaporation by the 1 MW/cm^2 beam will be close to adiabatic if its diameter is at least 1 mm. However, under conditions of fully developed evaporation, material is often found to be extracted by the beam much faster than expected from (5.19), due to ejection of melt, as will be discussed in Sect.5.2.3. Less energy is then invested in the blown-off material, and the adiabatic regime is reached earlier than the above estimates indicate.

The case of a penetrating source is more difficult to handle. If a steady state is ever reached, the surface velocity still obeys (5.19), but the relation between I_a and T_d is different from that for the surface source. Since the laser now heats a volume whereas evaporation only cools the surface, temperature maxima inside the material may develop. Approximate expressions for the temperature distributions have been obtained by *Dabby* and *Paek* [5.19], who found that the amount of superheating inside the dense phase (the surface was taken at T_{lv}) increases with the absorption length and with the ratio $\Delta H_{lv}/\Delta H(T_{lv})$. Such superheating beneath the surface favors volume evaporation and may result in explosive removal of the superheated region before it is fully vaporized. The surface then recedes in jumps rather than continuously. Expression (5.19) still holds on the average, although again with an enthalpy value on the RHS smaller than that for pure evaporation.

We have neglected heat losses by thermal radiation in the above analysis. The thermal flux emitted by a hot surface at temperature T is given by

$$\Phi^{rad} = \epsilon\sigma_{SB}(T - T_0)^4 \tag{5.21}$$

where $\epsilon < 1$ is the emissivity of the surface, and T_0 is the ambient temperature. Radiation losses by the condensed material can be allowed for in the same way as surface cooling by evaporation, but they are usually insignificant. At $T \simeq 3000$ K, Φ^{rad} is less than 300 W/cm^2, negligible compared to laser irradiances usually used in the evaporation regime.

The above analysis of heat flow in the condensed phase holds for pure near-equilibrium evaporation. However, evaporation by laser beams is usually neither pure – part of the blow-off may consist of melt – nor at near-equilibrium. To get a refined picture we must next consider some properties of the evolving vapor.

5.2.2 Vapor Expansion and Recoil

The vapor particles escaping from a hot surface have a Maxwellian velocity distribution corresponding to the surface temperature, but their velocity vectors all point away from the surface. This anisotropic velocity distribution is transformed into an isotropic one by collisions among the vapor particles. This happens within a few mean-free paths (typically of the order of a few μm) from the surface, a region known as the *Knudsen layer*. Some of the particles experience large-angle collisions and are scattered back to the surface. Beyond the Knudsen layer the vapor has reached a new internal equilibrium with a temperature different from the surface temperature. To obtain the properties of the vapor we treat the surface as a discontinuity and apply the conservation laws (5.10-12), with subscripts d denoting properties of the liquid surface and subscripts v those of the vapor just *beyond* the Knudsen layer. Furthermore, we treat the vapor as an ideal gas according to (5.13) and take the expansion velocity v_v equal to the local sound velocity, i.e.,

$$v_v = \sqrt{\gamma p_v / \rho_v} \tag{5.22}$$

where γ is the adiabatic index of the vapor. Thus we have 5 equations by which all properties of the vapor can be expressed in terms of, say, the surface temperature T_d. This calculation was performed by *Anisimov* [5.20] who, taking $\gamma = 5/3$, $A = 1$ and recognizing that $u \ll v_v$, found the following result

$$T_v \simeq 0.65 T_d \; ; \; \rho_v \simeq 0.31 \bar{\rho}(T_d) \tag{5.23}$$

where $\bar{\rho}(T) = \bar{p}(T) M / RT$ denotes the saturated vapor density at temperature T. This vapor is thus significantly cooler and less dense than the vapor in equilibrium with the surface. An analysis of the velocity redistribution in the Knudsen layer gave, under the same assumptions, that about 18% of all evaporating particles return to the surface. The net evaporation rate is thus somewhat smaller than the equilibrium value (5.6)

$$j_v \simeq 0.82 \bar{j}_v(T_d) \; . \tag{5.24}$$

The effective energy invested in the vapor exceeds the latent heat by the enthalpy and the kinetic energy of the vapor, or $\Delta H_{lv}^* = \Delta H_{lv} + H_v + M v_v^2 / 2$. Using (5.9) we find that

$$\Delta H_{lv}^* = \Delta H_{lv} + \frac{\gamma(\gamma + 1)}{2(\gamma - 1)} RT_d \simeq \Delta H_{lv} + 2.2 RT_d \; . \tag{5.25}$$

The solid line in Fig.5.3 represents the surface temperature in adiabatic evaporation with allowance for backflow, and with ΔH_{lv} replaced by the effective value ΔH_{lv}^*. Obviously the approximations made in (5.20) introduce only small errors as far as the surface temperature is concerned.

The simplest assumption about the further fate of the evolving vapor beyond the Knudsen layer is that of an adiabatic gas expansion. The pressure and temperature rapidly decrease away from the surface while the particle velocity increases [5.9]. If the vapor pressure significantly exceeds the ambient pressure, the flow velocity eventually becomes supersonic, while the pressure in the flow falls below the ambient pressure [5.21]. The supersonic low-pressure gas flow transforms into a subsonic flow at ambient pressure in a shock front which forms at a distance typically of the order of 1 cm from the target. Under stationary conditions (pulse duration long compared to the characteristic times of vapor expansion), the shock front is immobile with respect to the target and can be observed [5.22, 21]. However, it follows from the above numerical results that the vapor is supersaturated, i.e., that $p_v < \bar{p}(T_v)$, whenever the surface temperature T_d is smaller than about $\Delta H_{lv}/3R$, as is always the case in the present regime. Some of the vapor will therefore condense upon further expansion until a state of saturation is reached, and hence the expansion is not strictly adiabatic. A detailed discussion of vapor expansion and condensation can be found in the book by *Anisimov* et al. [5.24].

The dense phase is subject to recoil forces excerted by the evolving vapor. The actual pressure p_d is found from (5.23, 24) to be about half the saturated vapor pressure

$$p_d \simeq (1 + \gamma)p_v \simeq 0.54\bar{p}(T_d) . \tag{5.26}$$

The total force on the irradiated body is the integral of p_d over the irradiated area, while the momentum is the time integral of the force. For a pulse of duration t_p the total recoil momentum can be estimated from

$$\iiint p_d \, dx dy dt \simeq p_d \pi w^2 t_p \tag{5.27}$$

where w is the beam radius. Incidentally, this momentum exceeds that by the radiation pressure, equal to $F(1+R)t_p/c$, by many orders of magnitude.

The recoil momentum can be measured, e.g., by utilizing a target in the form of a small ballistic pendulum. Such measurements tend to reveal an optimum level of irradiance, characteristic of the material, where the momentum transfer per unit laser pulse energy is largest [5.2, 25]. That there should be an optimum is not surprising: At too low an irradiance a regime of adiabatic evaporation is never reached, and most of the pulse energy is spent in heating the condensed phase. As the irradiance increases, the surface temperature and the molar energy of the vapor also increase, reducing its efficiency in imparting momentum to the target (a larger number of less energetic particles carry more momentum than a lesser number of more energetic ones). In addition, more energy also goes into internal degrees of freedom of the vapor particles (dissociation, ionization) which do not contribute to the recoil. However, the dependence of momentum transfer on incident irradiance tends to reflect the behavior of the degree of absorption more than evaporation kinetics per se, as we shall discuss in the next section. We shall also consider further – and even more violent – recoil effects in Sects.5.3,4.

The primary target of the vapor recoil is, of course, the evaporating melt layer. The interaction affects the dynamics of evaporation and its response to transients in the irradiation [5.26]. Net forces onto the melt layer result only from pressure gradients, but lateral pressure gradients are always present, if only because a beam has a finite cross section. These forces invariably result in lateral displacement of liquid, which may range from mere changes in the surface topography of the irradiated material (Sect.2.3) to the formation of holes and deep welds.

5.2.3 Drilling, Welding, Cutting

Hole drilling is done with stationary beams, while welding and cutting consists essentially in moving a hole across the material by a scanned beam. Related techniques are turning and milling, in which the laser beam replaces the cutting tools in a lathe or a milling machine. Technical details and implementation data on a variety of laser machining processes have been given in [5.27]. Here we shall focus on the basic physical processes.

Hole Drilling

Let us consider hole drilling first. Drilling by pure evaporation is observed in sublimating materials or in metals at low irradiance. At higher irradiances, drilling velocities are often found to be much higher (perhaps by factors of 2 to 5) than can be accounted for by evaporation alone. The reason is that much of the material extracted leaves the hole as melt rather than as vapor. Surface-tension forces play a role in laser drilling of thin films for optical recording (Sect.2.3.2) [5.28], while melt extraction from deeper holes in bulk metals relies on the evaporation pressure. A simple model to help visualize this process is to think of the evaporating surface as a piston that exerts a pressure p_d onto the melt, squirting it out of the hole radially (Fig.5.4). The thickness of the melt layer can be estimated from (3.21) as

$$\delta_m \simeq \frac{\kappa}{u} \ln(T_{lv}/T_{sl}) . \tag{5.28}$$

In the following we treat the melt as a nonviscous, incompressible fluid. If, in a rough approximation, we replace the true pressure distribution by a "top hat" profile with a pressure p_d inside the beam radius w and ambient pressure p_0 outside, then the radial velocity of the melt follows from the volume work

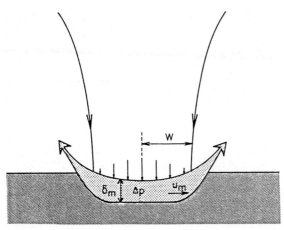

Fig.5.4. "Piston" mechanism of melt ejection by the evaporation recoil pressure. Small arrows symbolize the evaporation pressure distribution

$$u_m = \sqrt{2\Delta p/\rho_d} \qquad (5.29)$$

where $\Delta p \equiv p_d - p_0$ is the working overpressure. The liquid escapes through an opening area of the order of $2\pi w\delta_m$ along the circumference of the piston. The rate of melt ejection per unit beam area is found to be [5.29]

$$j_m \simeq \frac{1}{M}\left[(2\kappa/w)\ln\left(\frac{T_{lv}}{T_{sl}}\right)\right]^{1/2}\rho_d^{3/4}(2\Delta p)^{1/4} \qquad (5.30)$$

where the rate of material removal by evaporation alone has been neglected. The inverse relationship between the melt ejection rate and the beam radius makes the "piston" mechanism mainly relevant for focussed radiation. There is a threshold for melt ejection, determined by the surface tension of the melt, which must be overcome by the radial pressure force. Liquid metals have surface tensions of the order of 1 J/m^2 which gives threshold overpressures of the order of 10^4 Pa $\simeq 0.1$ atm for $w \simeq 0.1$ mm. However, for short pulses (1 μs or less) the minimum radial melt velocity required to carry the melt out of the beam spot while the pulse lasts (w/t_p) results in a significantly higher threshold than surface tension. Initial phases of melt ejection by a focused Gaussian Nd-laser pulse incident on a copper target are shown in Fig.5.5a: On the smooth molten surface of the first frame the evaporation pressure creates a shallow dimple which quickly develops into a doughnut-shaped billow expanding radially. The maximum radial velocity reached is close to 50 m/s, corresponding, according to (5.29), to a Δp of about 100 atm (the irradiance used here is, in fact, far above threshold, and the formation of a surface plasma, to be discussed in the next section, occurs between the third and the fourth frame of Fig.5.5a).

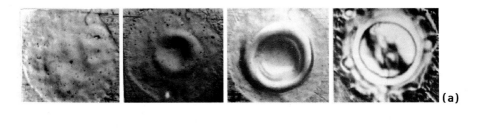
(a)

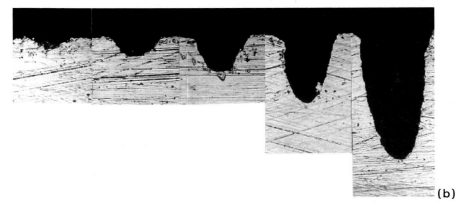
(b)

Fig.5.5. Illustration of melt ejection and hole drilling by 100 μm diameter Nd-laser pulses of rectangular temporal shape and variable duration: (**a**) early stages of melt ejection at 140 MW/cm^2 in copper, after 330, 380, 520 and 720 ns; (**b**) hole profiles in steel at 20 MW/cm^2, irradiated for 0.5, 1, 2, 5 and 10 μs (always from left to right)

Melt ejection results in a much larger velocity of the absorbing boundary than evaporation alone. The total velocity is the sum

$$u = V_d (j_v + j_m) \simeq V_d j_m \ . \tag{5.31}$$

The melt fraction in the total mass extracted by a focused laser beam in a metal can exceed 90 % at irradiances close to threshold. The effective energy content of the extracted material ($\simeq I_a / j_m$) then falls substantially below the value $\simeq [\Delta H(T_{lv}) + \Delta H_{lv}]$ for pure evaporation, and approaches the heat content of the melt. At higher irradiance the efficiency of the process gradually decreases, due to a smaller specific recoil as well as a smaller melt thickness (j_v scales linearly with the vapor pressure, j_m only with its fourth root). A series of micrographs illustrating hole drilling by a focused Nd-laser pulse are shown in Fig.5.5b. Note the "washed-out" appearance of the hole profiles, indicative of the action of laminar melt flow along the hole walls. Once the drilled hole grows deeper than about its radius, the irradiance required to sustain a steady flow of melt increases somewhat because of heat lost to the walls. The melt fraction decreases accordingly.

Melt can also be ejected from the irradiated zone by violent boiling. Boiling requires significant superheating of the melt, as well as pulses long enough to allow bubble nucleation. Melt ejection by boiling does not depend on focused radiation, but requires materials with a significant absorption length. The amount of melt ejected clearly depends on the superheating reached and thus on the heating rate, but quantitative estimates are difficult. The purity of the material and its content of dissolved gases would be expected to have a strong influence on the nucleation kinetics. Melt fractions of $60 \div 80\%$ of the total mass ejected have been observed in electrically exploded metal wires [5.30]. However, metals, due to their short absorption length, are unlikely candidates for laser-induced vapor explosions. Explosive ejection of material by laser irradiation was reported in the case of ceramics by *Gagliano* and *Paek* [5.31]. Using irradiances of a few 10^7 W/cm^2, they found the amount of material ejection to vary strongly with irradiance and observed material ejection to continue even after the end of the pulse. Hole drilling by the explosive mechanism is, in general, more difficult to control than drilling by melt ejection, and the holes tend to be less regularly shaped [5.32]. Ionization of the vapor may occur before the nucleation temperature is reached in the liquid. The associated increase in pressure tends to suppress boiling by reducing the amount of superheating in the melt. Drilling then occurs by normal surface evaporation. Even highly transparent materials like sapphire can be drilled very well by intense pulses that cause optical breakdown of the material near the beginning of the pulse [5.33].

Penetration Welding

Let us now consider penetration welding. It is typically done by means of CW CO_2 lasers with powers of the order of several kW. Welding depths in steel can reach 1 cm or more, depending on beam power and scan speed, and are obviously unrelated to the light absorption length. Evidence shows that the focused beam forms a narrow cylindrical vapor cavity in the melt which "traps" the beam and moves through the workpiece as the beam is scanned. Figure 5.6 exhibts a schematic of the vapor cavity and the associated melt volume in a homogeneous slab of metal. The actual welding occurs in the solidifying melt puddle trailing the vapor cavity. The cavity formed by a beam scanned at constant velocity can be regarded as a hole at steady state: while for a stationary beam the hole grows indefinitely (in principle), the material extracted by the scanned beam is constantly replenished as the beam moves forward. However, a stable steady state is only found within certain limits of the parameters beam power, spot radius, and scan speed.

Quantitative models of the complex heat and mass flows in penetration welding are available [5.34]. Here we shall follow a somewhat simplified analytical treatment by *Klemens* [5.35] which well illustrates the essential

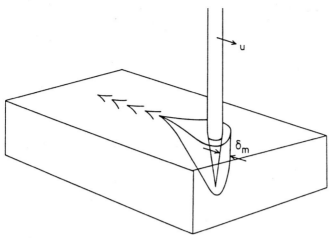

Fig.5.6. Schematic of vapor cavity and melt zone during penetration welding with a scanned high-power beam

features. He took the surface of the cavity to be at T_{lv} and treated the melt flow in a horizontal plane. The width δ_m of the melt layer in front of the cavity (Fig.5.6) is estimated by requiring that the heat flux across the layer, $K(T_{lv}-T_{sl})/\delta_m$, heats and melts material at exactly the rate $u[\Delta H(T_{sl}) + \Delta H_{sl}]/V$ required for propagation at the scan speed u. This gives

$$\delta_m = \frac{\kappa}{u} \frac{T_{lv} - T_{sl}}{T_{lv} - \Delta H_{lv}/c_p} . \tag{5.32}$$

The melt is forced to flow around the cavity by the evaporation pressure at its front. *Klemens* assumed that a fraction β of the material intercepted by the cavity is evaporated, hence

$$\rho_v v_v = \beta \rho_d u . \tag{5.33}$$

The momentum transported by the flow of vapor creates the excess pressure

$$\Delta p = \rho_v v_v^2 = \beta \rho_d u v_v . \tag{5.34}$$

The melt again acquires a velocity related to the overpressure by $u_m = (2\Delta p/\rho_d)^{1/2}$. The melt volume flowing around the cavity on each side per unit time and unit cavity length, $u_m \delta_m$, is a fraction $(1-\beta)$ of the total volume, uw, swept out by the corresponding half-cavity of radius w. Combining this requirement with (5.32-34) yields

$$\beta \simeq \frac{wu}{\kappa} \frac{T_{lv} - \Delta H_{lv}/c_p}{T_{lv} - T_{sl}} \sqrt{\rho_v/2\rho_d} . \tag{5.35}$$

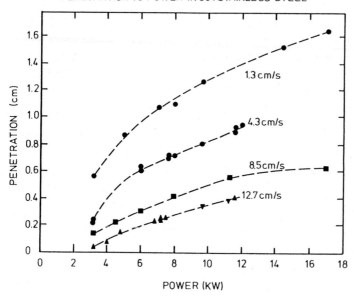

Fig.5.7. Weld penetration in 304 stainless steel as a function of CO_2-laser power at various scan speeds [5.36]

Since ρ_v/ρ_d is of the order of 10^{-4} near T_{lv}, β is small. Choosing typical parameters (steel, w = 1 mm, u = 3 cm/s), *Klemens* found $\beta = 0.02$, $\Delta p = 16$ Pa and $u_m = 4$ m/s. The absorbed beam power to create a stable 1 cm long cavity at this scanning speed is obtained as about 4 kW, with 3 kW spent on melting and only some 100 W on evaporation (the remainder is lost through heat conduction into the solid). Given that the cavity can be expected to trap most of the incident beam power, the process is quite efficient.

In reality the processes during high-power laser penetration welding are quite a bit more complex that the above analysis suggests. The pressure inside the cavity varies over its cross section as well as over its depth. *Klemens* showed that, as a result of vapor viscosity and surface tension, actual cavity shapes are cone-like with constricted exit holes, rather than cylindrical. Moreover, the hot vapor inside the cavity becomes partially ionized and absorbing as the beam power increases. This, together with the increasing vapor fraction β, causes the penetration depth to saturate as a function of beam power. This is reflected in Fig.5.7, which shows measured penetration depths in steel as a function of beam power for various scan speeds [5.36].

Cutting

If the length of the vapor cavity reaches or exceeds the thickness of the irradiated metal slab, it can be used for cutting. However, cutting sheet metal by laser beams is usually – and far more efficiently – done with a jet of gas assisting the laser [5.36]. The gas jet, rather than evaporation pressure, provides the momentum to expel the melt, so the laser must only melt the material. Gas flows with Mach numbers around 0.2 appear to be most efficient in removing the melt [5.37]. In the case of steel and other reactive metals oxygen gas is often used. Exothermic oxidation reactions then contribute to the power available for cutting [5.38]. Although the gas flow also cools the interaction zone, the cutting speed for a given laser power can be doubled by an oxygen jet. Sheet metals can be cut at acceptable speeds with beam powers as small as 200 W. Apart from metals, a large variety of industrial materials, ranging from ceramics to leather, are nowadays routinely and successfully cut with scanned CO_2-laser beams.

5.3 Absorption Waves

We shall now move on to irradiances at which the interaction of the beam with the evolving vapor plays a dominant role. In this regime, the vapor becomes ionized and absorbs part or all of the incident radiation. The energy is converted into internal energy of the plasma, radiated away as thermal radiation or consumed in hydrodynamic motion. The properties of the evaporating surface become less important and may be treated as mere boundary conditions for the processes in the hot ionized vapor.

Since the density and the temperature of the vapor decrease away from the target, the plasma is most likely to form close to the evaporating surface. The temperature and the degree of optical thickness reached in the vapor depend on the incident irradiance. The irradiance received by the condensed surface is accordingly reduced. Once an absorbing gas plasma has formed, an intriguing phenomenon is observed: The plasma cloud rapidly expands away from the surface, but remains essentially confined to the light channel formed by the beam. This has little to do with ordinary gas expansion, but results from the dynamics of interaction of the light with the plasma. The propagating plasma is generally referred to as a Laser-Supported Absorption Wave (LSAW). Figure 5.8 exhibits a series of high-speed photographs illustrating ignition and expansion of LSAWs upon irradiation of (a) an aluminium and (b) an alumina target with CO_2-laser pulses [5.39]. The dynamics of formation and propagation of such LSAWs significantly influences the interaction of the beam with the target.

(a)

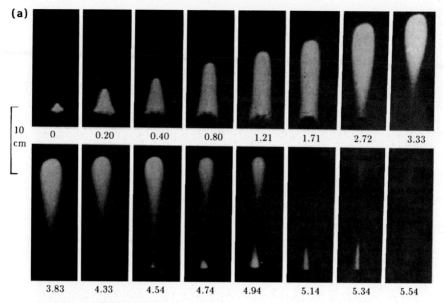

| 10 cm | 0 | 0.20 | 0.40 | 0.80 | 1.21 | 1.71 | 2.72 | 3.33 |

| 3.83 | 4.33 | 4.54 | 4.74 | 4.94 | 5.14 | 5.34 | 5.54 |

Fig 5.8. See caption on the opposite page

LSAWs come in widely varied shapes and features, depending on experimental conditions such as irradiance, wavelength, focal-spot size, target material and ambient atmosphere. Although, in practice, the distinction is not always easy to make, LSAWs can be roughly divided into two classes, depending on whether they propagate at subsonic or at supersonic speed with respect to the gas (vapor or ambient atmosphere). The two classes further differ in the range of irradiance in which they are typically observed, as well as in the optical density and internal energy reached by the plasma. We shall first consider the weakly absorbing subsonic variety of which Fig.5.8 displays two examples. Such plasmas are generally known as Laser-Supported Combustion Waves (LSCWs).

5.3.1 Laser-Supported Combustion Waves

As the name suggests, there is a basic similarity between a LSCW and a chemical combustion wave which propagates in a combustible gas mixture. The source of energy driving the latter is an exothermic chemical reaction with a reaction rate that typically follows an Arrhenius law. The chemical combustion wave propagates as layers of cold gas adjacent to the hot reaction zone are heated by conduction or thermal radiation until they start burning and producing heat themselves. In the LSCW the role of the chemical reaction is played by light absorption, which, according to (5.15), also behaves in an Arrhenius-like manner. Propagation is limited to the channel

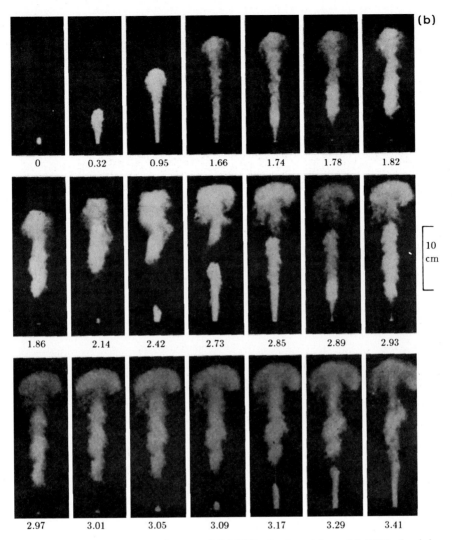

Fig.5.8. High-speed camera frames of LSAW's developed from (a) 2024 aluminium alloy and (b) alumina targets, irradiated by 5 ms, 1.5 MW/cm^2 CO_2-laser pulses (incident from above). Several decoupling and reignition events can be seen. Numbers below frames give the time in ms [5.39]

formed by the laser beam, but it relies on the same mechanisms of heat transport as the chemical combustion. In contrast to the supersonic propagation modes to be considered later, an LSCW tends to be optically thin to the laser radiation, i.e., it absorbs only a fraction of the beam flux. For this reason LSCWs are often observed to propagate away from the point of ignition in two directions simultaneously – towards as well as away from the

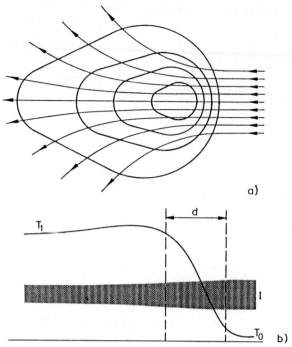

Fig.5.9. (a) Flow lines (arrows) and isothermal contours of a laser-supported combustion wave; (b) schematic of gas heating in the LSCW

laser beam – provided no target blocks the latter path. Otherwise a plasma stationary with respect to the target is formed.

The theory of LSCW propagation was formulated by *Raizer* [5.40] and has since been refined by a number of workers. The basic structure of an LSCW is shown in Fig.5.9a. In the wave's own frame of reference, cold gas enters at the front, is heated by conduction and light absorption and leaves the wave at the rear end as well as laterally. The gas within the wave gains energy from the laser beam and loses energy by radiation and heat conduction to the surrounding gas. Mathematical modelling of the LSCW is complicated because the problem is three-dimensional, and because the absorption and radiation properties of weakly absorbing plasmas depend on the gas composition, the temperature and the wavelength in a rather detailed fashion. Let us consider here a simplified one-dimensional problem, treated by *Boni* and *Su* [5.41], which well illustrates the qualitative features of LSCWs.

We regard the LSCW as a disk of thickness d, within which gas is heated from a temperature T_0 to T_1 (Fig.5.9b). The energy balance of the wave may be written as

$$K^{\text{eff}}(T_1 - T_0)/d = d(I\alpha - J^{\text{loss}}) \tag{5.36}$$

where α is the absorption coefficient at the laser frequency, J^{loss} is the volumetric energy loss rate of the plasma due to conduction and thermal radiation, and K^{eff} is an effective thermal conductivity allowing for diffusive as well as radiative energy transport

$$K^{eff} = K + K^{rad} .\qquad(5.37)$$

For radiative transport to be treated in terms of a conductivity, the medium must be optically thick, i.e., its linear dimensions must exceed the absorption length for the important part of its thermal radiation. Furthermore, there must be local equilibrium between the medium and the radiation. The blackbody radiation density at any point is then characteristic of the local temperature. This is the case provided the temperature varies negligibly over a distance of one radiation mean-free path [5.10]. Under these conditions the radiative conductivity is

$$K^{rad} = \frac{16}{3}\sigma_{SB}\,\ell_R\,T^3 \qquad(5.38)$$

where ℓ_R is an averaged radiation mean free path known as the Rossland mean-free path,

$$\ell_R = \frac{15}{4\pi^4}\int_0^\infty \frac{1}{\alpha(\omega)}\frac{(\hbar\omega/kT)^4 e^{-\hbar\omega/kT}}{(1-e^{-\hbar\omega/kT})^3}\,d\!\left(\frac{\hbar\omega}{kT}\right). \qquad(5.39)$$

The weighting function inside the integral is maximum at $\hbar\omega \simeq 4kT$, indicating that the high-energy photons are dominant in the energy transfer process. The condition of thermal equilibrium is usually satisfied for LSCWs, whereas the condition of optical thickness only holds for the vacuum-ultraviolet part of the thermal emission. In their model, *Boni* and *Su* approximated the true absorption spectrum of the plasma by a two-band spectrum with a large constant absorption coefficient above some limiting photon energy (10.9 eV for air) and a negligible one below. The high-frequency part of the thermal radiation thus contributes to the energy transport. The low-frequency part which dominates at the temperatures typical for LSCWs (of the order of $1\div 2\,eV$), has an absorption length exceeding typical plasma dimensions and is treated as a loss. The mass of gas heated in the wave to the final temperature T_1 is now given by

$$\rho_0\frac{u_0}{d} = \frac{M(\alpha I - J^{loss})}{c_p(T_1 - T_0)} \qquad(5.40)$$

where u_0 is the propagation velocity of the LSCW. Eliminating the wave thickness d from (5.36, 40) yields for the wave velocity

$$u_0 = \frac{MK^{eff}}{\rho_0 c_p} \sqrt{\frac{\alpha I}{K^{eff}(T_1 - T_0)}\left(1 - \frac{J^{loss}}{\alpha I}\right)} \quad . \tag{5.41}$$

The LSCW velocity thus scales essentially with the square root of the irradiance and vanishes at a critical irradiance equal to J^{loss}/α, at which the laser beam just balances the energy loss rate. The LSCW then becomes stationary with respect to the surrounding gas [5.42]. Such stationary plasmas, known as plasmotrons, have been sustained (after spark ignition) with CW CO_2 lasers at powers below 100 W in Ar or Xe gas at one atmosphere [5.43]. At irradiances well above the critical value, LSCW velocities in the range of a few 10 to a few 100 m/s are found to be typical [5.44].

The apparent simplicity of (5.41) is delusive: A realistic evaluation of K^{eff} and J^{loss}, which both depend sensitively on the nature and temperature of the gas as well as on the geometry of the problem, is quite complicated. Allowance for just two frequency bands is, of course, an oversimplification. To obtain more than just qualitative agreement with experiment, a numerical treatment of the hydrodynamic and radiation-transport problems is required. In the case of LSCWs formed in a target vapor, one must also allow for the fact that the wave propagates in a medium which is already hot and in motion [5.45]. The range of irradiance where the LSCW model applies extends, roughly speaking, from tens of kW/cm^2 to tens of MW/cm^2, and covers the range where intense evaporation of targets occurs within the pulse duration. However, these fluxes are too low to ionize cold gases and hence vapor-LSCWs, as a rule, do not propagate beyond the region occupied by hot vapor.

Let us now return to the evaporating target and consider how LSCWs interfere with the beam-material interaction. One might think of an LSCW which extracts energy from the beam and dissipates it by expansion or radiation into 4π, as a mere parasite in the beam-target interaction. Experience shows, however, that at least under certain circumstances LSCWs can actually enhance the amount of beam energy deposited in a solid.

5.3.2 Plasma-Enhanced Coupling

As mentioned in Sect.2.3.3, it is well established that the fraction of beam flux absorbed in metals increases above a certain threshold irradiance. The increase tends to be particularly large (up to a factor of ten or more) for infrared lasers which are poorly absorbed by cold metals. There is clear evidence that such an effect can be produced by formation of an LSCW in the evolving vapor.

How does this phenomenon of plasma-enhanced coupling work? A typical sequence of events might be as follows. First, material is heated to the point of evaporation by normal absorption. The vapor close to the surface is partially ionized and begins itself to absorb significantly [5.13]. A stationary plasma layer close to the evaporating surface forms, often coincident with initiation of a propagating LSCW moving up-beam. The temperature of the near-surface plasma is typically in the range of $1 \div 3$ eV [5.46]. Energy from the plasma can now be transferred to the dense phase by any of three mechanisms: (i) normal electron heat conduction, (ii) short-wavelength thermal plasma radiation which is efficiently absorbed by the metal surface, and (iii) condensation of vapor forced back to the surface by the plasma pressure [5.47]. These mechanisms provide an additional heat flux to the dense material, which may or may not exceed the loss of light flux due to plasma absorption. One would speak of plasma-enhanced coupling in the former and plasma shielding in the latter case. Most experiments show that the enhanced coupling phenomenon is strongest right at its threshold and decreases at higher irradiance [5.48], suggesting that the plasma tends to decouple from the surface. The energy transferred from the plasma to the target is often found to be distributed over an area significantly larger than that of the optical beam spot, due to lateral expansion of the plasma [5.49].

Let us consider a slightly simple-minded model of the process, sketched in Fig.5.10, in order to illustrate some of its features (a somewhat more complete model, yielding almost identical conclusions, has been discussed by *Nielsen* [5.50]). The plasma is represented as a disc of thickness d. The dense target absorbs a fraction $(1-R)$ of the irradiance I_1 transmitted through the plasma. A fraction β of the light flux $(I-I_1)$ ab-

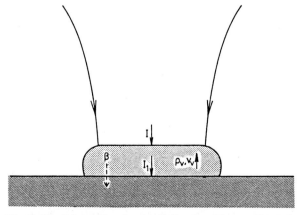

Fig.5.10. Schematic of a partially absorbing surface plasma with energy transfer to the dense phase

sorbed by the plasma is assumed to be transferred to the target by any of the mechanisms mentioned above. The total flux deposited in the target is written as

$$AI \equiv (1-R)I_1 + \beta(I-I_1) = \rho_v v_v \Delta H_{lv}^* / M . \tag{5.42}$$

The left-hand identity serves to define the effective beam-target coupling coefficient A, whereas the right-hand equation expresses energy conservation at the target-plasma interface. The amount of absorption in the plasma depends on its density and its temperature. Of these, the density varies strongly with the flux absorbed in the target, whereas the temperature does not. In a weakly ionized plasma we have $\alpha \propto \rho_v$, (5.15), whereas in a fully ionized one, which we shall assume here, we have $\alpha \propto \rho_v^2$, (5.16). The total absorption in the plasma will also depend somewhat on the kinetic energy of the vapor which affects the length over which the density decreases, but the dependence is not strong and we just neglect it here. Thus we can write for the flux transmitted to the target

$$I_1 = I e^{-B_0 \rho_v^2} \tag{5.43}$$

where, for a given laser wavelength, B_0 is a constant; according to (5.16), B_0 scales like λ^2. Inserting (5.43) into (5.42) yields an implicit equation for A, i.e.,

$$A - (1-R-\beta)e^{-B_1 A^2} - \beta = 0 \tag{5.44}$$

where $B_1 = B_0 (I/v_v \Delta H_{lv}^*)^2$. Figure 5.11 plots the effective coupling coefficient calculated from (5.44) as a function of $B_1 \propto I^2$ for $(1-R) = 0.05$ and for various values of β. Note that there is an abrupt threshold for plasma-enhanced coupling if β is large (even though no threshold is built into the assumptions, the vapor always being taken to be fully ionized), while the transition becomes gradual at smaller β. The transition in all cases occurs while the plasma is still transmitting most of the laser light. The case $\beta = 0$ leads to reduced coupling (shielding) at larger irradiance.

Good thermal coupling (large β) evidently requires the plasma to stay in close proximity to the surface. Any decoupling (drop in β) would immediately reduce the rate of evaporation and hence the density of the vapor and the absorbing power of the plasma. The close circular relationship between the plasma-target energy-transfer rate, the evaporation rate and the plasma density may lead either to a self-balanced steady state (corresponding to a non-propagating LSCW) or to oscillatory behavior (repeated ignition and decoupling of propagating LSCWs), always assuming the incident irradiance to be constant. The latter behavior is evident in Fig. 5.8 where decou-

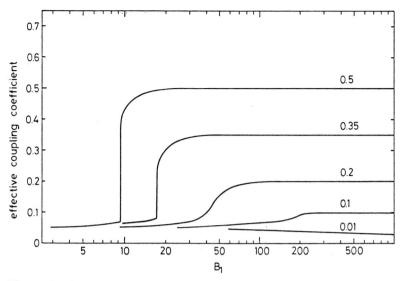

Fig.5.11. Effective coupling coefficient calculated from (5.44) for various values of the energy transfer factor β. B_1 is proportional to the square of the incident irradiance

pling occurs, e.g., in series (a), frames $2.72 \div 4.33$ ms, and reignition is visible at 4.54 ms (there is no correlation of these events with features of the incident pulse envelope, as given by *Stegman* et al. [5.39]). Repeated decoupling and reignition is also obvious in series (b) of Fig.5.8. The target surface temperature monitored under interaction showed strong maxima during periods of coupling and minima during periods of decoupling.

The behavior of the LSCW, as well as the resulting degree of coupling or shielding, apparently depends on a variety of parameters – the laser flux and wavelength, the target material as well as the ionization potential and pressure of the ambient gas [5.39, 51]. Large plasma-enhanced beam coupling to metals (up to $30 \div 50\%$ of incident beam power absorbed in the condensed phase) have only been reported for infrared laser beams, notably for Nd lasers [5.52] and for CO_2 lasers [5.53].

With further increasing irradiance the temperature, pressure and velocity of the absorption wave increase. At the same time the wave becomes more absorbing and consumes a larger fraction of the beam flux. Compression of the gas engulfed by the LSAW contributes more and more to preheating and ionization. Eventually compression, rather than heat conduction, becomes the dominating propagation mechanism. The velocity in this regime turns out to be supersonic with respect to the gas ahead of the wave, hence the wave is a shock wave. Waves of this kind are known as Laser-Supported Detonation Waves (LSDWs).

5.3.3 Laser-Supported Detonation Waves

The absorption wave due to a laser-supported detonation is, in contrast to LSCWs, not confined to the vapor but can form and propagate in cold gases, too. In fact, LSDWs are typically observed to form ahead of a target before the latter has even started to evaporate, due to surface-mediated optical breakdown of the ambient gas, as discussed in Sect.5.1.4. Breakdown plasmas tend to be strongly absorbing and can, once ignited, readily be sustained and propagated as LSDWs. The name of LSDW alludes to its similarity to a strong chemical explosion which propagates by shock compression of a combustible mixture. In the LSDW light absorption plays the role of the chemical reaction. Typical plasma temperatures reached in LSDWs are of the order of 10 eV, at which most gases are multiply ionized. Absorption lengths for infrared-laser beams are typically no more than a few tens of μm, small compared to usual beam dimensions.

This being so, the main characteristics of an LSDW – temperature, pressure, and propagation velocity – can be found by neglecting its internal structure altogether. The wave is treated as a mere hydrodynamic discontinuity at which the incident light flux is deposited (Fig.5.12). To find its characteristics we apply the conservation laws (5.10-12). Denoting quantities in front of and behind the wave with subscripts 0 and 1, respectively, and neglecting the pressure and the enthalpy of the undisturbed gas we have

$$\rho_0 v_0 = \rho_1 v_1 \, , \tag{5.45}$$

$$\rho_0 v_0^2 = p_1 + \rho_1 v_1^2 \, , \tag{5.46}$$

$$v_0^2/2 = H_1/M + v_1^2/2 - I/\rho_1 v_1 \, . \tag{5.47}$$

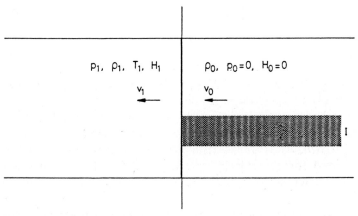

Fig.5.12. Hydrodynamic variables for calculating the characteristics of a laser-supported detonation wave. Subscript 0 identifies the undisturbed ambient gas, subscrip 1 the hot gas behind the wave (moving frame of coordinates)

Here the laser irradiance is assumed to be fully absorbed at the discontinuity and thus appears as additional enthalpy of the gas behind the wave. Further, we take the velocity of the gas behind the shock wave to be at its maximum possible value – the local sound velocity

$$v_1 = (\gamma p_1 / \rho_1)^{1/2} . \tag{5.48}$$

Together with the equation of state (5.9) for H_1 we have 5 equations which allow us to eliminate all quantities pertaining to the gas behind the wave, and to express the wave velocity in terms of the irradiance. This yields for the wave velocity $-v_0 \equiv u$ [5.54]

$$u = [2(\gamma^2 - 1)I/\rho_0]^{1/3} . \tag{5.49}$$

Substituting, for illustration, $I = 10^{10}$ W/cm^2 = 10^{14} W/m^2, $\rho_0 = 1.3$ kg/m^3 (air at STP) and $\gamma = 4/3$ into (5.49) yields $u \simeq 50$ km/s. The gas temperature can be estimated from the internal energy behind the front

$$U_1 = \frac{H_1}{\gamma} = \frac{Mu^2\gamma}{(\gamma^2 - 1)(\gamma + 1)} . \tag{5.50}$$

In the above example we get $U_1/M \simeq 1.8 \; 10^9$ J/kg, corresponding to an equilibrium gas temperature of about 120.000 K. The compression has the value

$$\rho_1/\rho_0 = 1 + 1/\gamma \tag{5.51}$$

which, in our example, is 7/4. Finally, the pressure behind the wave is found to be

$$p_1 = \frac{\rho_0 u^2}{\gamma + 1} \tag{5.52}$$

or approximately $1.4 \cdot 10^9$ Pa $\simeq 14000$ atm. The LSDW, as these numbers illustrate, is certainly a rather violent phenomenon.

The above calculation neglects lateral expansion of the gas, which represents a power loss to the LSDW and reduces its temperature and velocity. According to *Raizer* [5.54], the loss to lateral gas expansion can be approximately allowed for by replacing the incident flux I in (5.49) by a reduced flux $I/(1 + d/w)$, where $d \propto \alpha^{-1}$ is the thickness of the wave and w is the beam radius.

Apart from the detonation mechanism, other propagation modes for supersonic LSAWs have been considered. One of them is the *breakdown wave* [5.54] which applies for light fluxes exceeding the breakdown threshold in cold gas (without a target). Propagation of the breakdown wave relies on the time dependence of avalanche breakdown: The avalanche first develops in the region of highest flux (usually the focal point), and somewhat later at points of lower flux and longer avalanche build-up time. The velocity of this "wave" is inversely proportional to the beam-aperature angle and hence can be arbitrarly large. Since plasmas in front of absorbing targets are formed at irradiances much below those required for breakdown in cold gases, the breakdown wave is of little relevance for laser-target interactions. Yet other possible LSAW propagation modes rely on the far-ultraviolet thermal radiation of the hot plasma ($\hbar\omega$ of the order of tens of eV), for which the plasma itself is transparent, but which is strongly absorbed in a cold ambient gas [5.54, 55]. The experimental distinction between an LSDW and the radiation-assisted LSAW modes (which have a far more complicated mathematical description) is not obvious, however. We shall henceforth refer to a supersonic LSAW as an LSDW.

Let us now return to the evaporating target and consider how LSDWs interfere with the beam-material interaction. Their role turns out to be quite different from that played by LSCWs. Enhanced coupling may be present initially, but it is quickly replaced by complete shielding of the target as the strongly absorbing plasma propagates away from the surface.

5.3.4 Effects of LSDWs on the Beam-Material Interaction

Both the coupling and the shielding phases of target-mediated LSDW formation have been studied mainly in connection with high-power CO_2 lasers for which the threshold is about 10^7 W/cm^2. The period of enhanced coupling after ignition of an LSDW lasts typically less than one μs and is dominated by mechanical rather than thermal energy transfer.

The LSDW, as shown above, creates a pressure of hundreds or thousands of atmospheres. The gas expands behind the wave and the particle velocity must vanish on the surface of the target. Assuming the expansion to be isentropic and one-dimensional, *Pirri* [5.56] obtained for the actual pressure acting on the target surface

$$p_2 = p_1 \left[(\gamma + 1)/2\gamma\right]^{2\gamma/(\gamma-1)} \tag{5.53}$$

where p_1 is given by (5.52). This should be valid until the LSDW has moved away from the target further than about its diameter. The pressure increases with increasing irradiance, but the time during which the pressure acts on the target decreases because the LSDW moves faster. The total momentum, in the one-dimensional limit, turns out to be proportional to the total pulse energy and independent of irradiance (provided the latter is well above the threshold for LSDW ignition). A pictorial impression of this regime is given

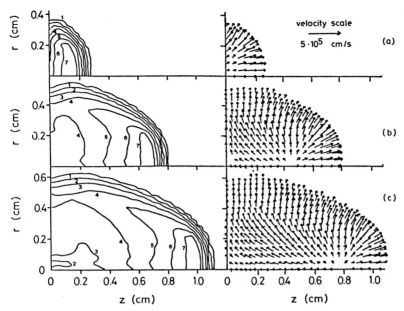

Fig.5.13. Calculated pressure contours and flow fields in a LSDW sustained by a 10 MW/cm² CO_2-laser beam: (a) after 0.66 μs, (b) after 1.63 μs, (c) after 2.19 μs. Pressure contour scale (in 10^6 erg/cm³): 1:1.8, 2:3.6, 3:7.2, 4:14.4, 5:28.8, 6:57.6, 7:115. [5.57]

by Fig.5.13, which shows numerically determined pressure contours and flow vectors in an LSDW formed at a surface by a 10 MW/cm² CO_2-laser pulse [5.57]. Equations (5.49, 52, 53) yield $p_2 \simeq 35$ atm, in fair agreement with the surface pressure predicted by Fig.5.13 while the gas motion is one-dimensional. Direct measurements by means of piezoresistive gauges [5.58] also confirmed the peak pressures predicted by theory.

   LSDW-related pressure pulses produce stress waves in the target material which can lead to deformation or fracture. Effects in various materials have been investigated by a number of researchers [5.59]. The impulses delivered to reflecting materials by infrared lasers through LSDW ignition by far exceed those expected from normal evaporation at the same irradiance (even stronger impulses are produced in the regimes to be considered in the next section). However, the period of LSDW-enhanced coupling usually lasts only for a small fraction of the pulse duration. As soon as the LSDW is decoupled, the target is completely shielded and receives no further energy from the laser beam while the LSDW lasts. The fraction of pulse energy transmitted to the target thus depends on the lifetime of the LSDW.

   In focused beams the flux I received by an LSDW moving away from the focal region decreases in proportion to the inverse square of the aperture angle and the distance travelled by the wave. Eventually the flux inci-

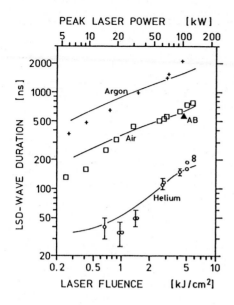

Fig.5.14. Lifetime of the LSDW created by a TEA CO_2-laser pulse (500 ns spike, followed by a 4 μs tail) focused by a f/5 lens onto a target in various ambient gases. The point AB indicates the air-breakdown threshold without a target. Solid lines were calculated from LSDW theory. [5.61]

dent on the LSDW is insufficient to further sustain its losses, the wave decays and light is again transmitted to the target. The "lifetime" of the LSDW thus depends on the aperture angle of the light cone. The minimum "keep-alive" flux is found to be somewhat smaller than that required to ignite the wave [5.60, 61]. The expanding shock wave leads to rarefaction of the gas in front of the target which may persist for tens of μs and which tends to suppress reignition of subsequent LSDWs [5.23]. The fraction of pulse energy lost to shielding can be minimized by using large-aperture focusing optics and ambient gases with a large sound velocity [5.61]. LSDW lifetimes in various gases as a function of laser fluence are shown in Fig. 5.14. The dependence of the hole-drilling efficiency on pulse duration and irradiance in this regime has also been discussed by *Hamilton* and *Pashby* [5.62].

5.4 Phenomena at Very High Irradiance

This section is devoted to regimes accessible only with light fluxes exceeding some 10^9 or 10^{10} W/cm^2. The interest in this area has been stimulated mainly by the prospect of producing plasmas hot and dense enough to achieve thermonuclear fusion. Other potential appliactions include the use of plasmas as pulsed sources of fast ions or of X rays. Stimulated X-ray emission at wavelength as short as 4 nm has been realized utilizing sophisticated geometrical arrangements of high-density laser-produced plasmas [5.63].

The regime under consideration, while fascinating, is somewhat peripheral to the main theme of this monograph, so we will merely sketch the main ideas, in order to provide, as it were, a glance across the fence delimiting regimes of current interest in laser material processing.

In the previous section we have been concerned with gaseous plasmas with an optical density increasing with irradiance. Once matter is fully ionized this trend is reversed since the degree of ionization cannot increase any further. The absorption coefficient of a fully ionized plasma, given in (5.16), scales like $\rho^2 T^{-3/2}$, hence a sufficiently hot plasma will be transparent to the laser radiation. Consequently, radiation is again transmitted to the dense phase. The temperature at which this happens depends on the atomic number of the material. Most of the relevant work has been concerned with the lightest elements, which are of interest in fusion, and where full ionization is relatively easy to achieve.

At irradiances typical for the present regime, even nonmetals can be considered strongly absorbing, since breakdown will occur within a small fraction of the pulse duration. At the highest irradiances (10^{15} W/cm^2 or more) ionization occurs by multiphoton absorption, which becomes equivalent to the tunnel effect, essentially within one light cycle. The specific energy acquired by the material exceeds the heat of evaporation by many times, and even the dense phase can be treated as an ideal gas. However, strong nonequilibrium between the electrons and the ions will prevail while such a pulse lasts.

We are thus confronted with the problem of a dense, strongly absorbing material, in the first few tens of nm of which energy at a rate of perhaps 10^{20} W/cm^3 is liberated. Part of this energy, once randomized, is conducted into the bulk of the material, while part is converted into directed kinetic energy by thermal expansion of the heated layer. It has been customary to distinguish two regimes in which most experiments have been done: First, a regime characteristic of ns (Q-switch laser) pulses, which is dominated by the expansion and ablation of material, and second, a regime, characteristic of ps (mode-locked laser) pulses, in which heat conduction dominates, as hydrodynamic motion during the pulse duration is negligible.

Let us first consider the former regime. The main difference to that considered in Sect.5.2 is that the thermal pressure of the heated layer, orders of magnitude greater, is now sufficient to cause significant compression of the underlying target material. In the simplest case we have to deal with three different zones (Fig.5.15): The undisturbed solid (denoted with the subscript 0), the compressed layer (subscript 1) and the vapor (subscript 2). We can apply the conservation laws (5.10-12) to the boundaries between the three regions, but for a full description, as before, additional assumptions must be made. Various models have been considered which differ mainly in the treatment of light absorption in regions 1 and 2. A model appropriate for intermediate fluxes (between some 10^9 and 10^{14} W/cm^2 for

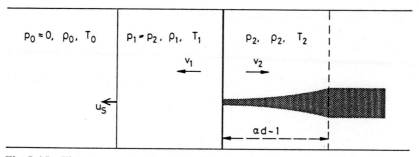

Fig.5.15. Three-zone model of irradiated material in the self-regulating plasma regime. Subscripts O, 1, 2 identify the undisturbed solid, the compressed layer and the expanding vapor plasma, respectively. Laboratory coordinates

light targets) is based on the idea, first proposed by *Krokhin* [5.64], of a self-regulating plasma.

5.4.1 The Self-Regulating Plasma

The basic idea is simple and has been hinted at in Sect.5.3: Absorption in the gaseous plasma in region 2 will decrease the rate of evaporation and hence the plasma density. The plasma is thus expected to transmit just enough radiation to the dense surface in order to sustain itself. A very simple but intuitively appealing model is that of a "plasma between walls" [5.64]. The plasma is thought to have a constant thickness d. The plasma temperature due to absorption of laser radiation (with constant irradiance I) then increases according to

$$\frac{\partial T_2}{\partial t} = MI \frac{1 - e^{-\alpha d}}{d\rho_2 c_p} \tag{5.54}$$

where α is the absorption coefficient at the laser wavelength. From (5.16), it can be written as

$$\alpha = B_2 \rho_2{}^2 T_2{}^{-3/2} \tag{5.55}$$

with a constant B_2. The plasma density, in turn, depends on the irradiance transmitted

$$\frac{\partial \rho_2}{\partial t} = \frac{I e^{-\alpha d}}{d c_p T_2} . \tag{5.56}$$

The system (5.54-56) has the solution

$$T_2 = (\alpha B_2)^{2/7} (ItM/c_p)^{4/7} ;$$

$$\rho_2 = (\alpha B_2{}^2)^{-1/7} (ItM/c_p d^2)^{3/7} \tag{5.57}$$

which is easily verified by insertion. The optical thickness of the plasma, αd, turns out to be equal to $\ln(7/3) \simeq 0.85$, independent of time and irradiance. This means that 3/7 of the pulse energy is spent in evaporation and 4/7 in plasma heating. This simple model demonstrates, if nothing else, that effective self-regulation of the plasma is possible.

A somewhat more realistic model, also based on the idea of self-regulation of the plasma but allowing for shock compression of the solid was developed by *Caruso* and *Gratton* [5.65] (Fig.5.15). The particle velocity in the plasma region, assumed to contain ions of charge Z only, is taken as the isothermal sound velocity

$$v_2 = \sqrt{(1 + Z)RT_2/M} \tag{5.58}$$

and the physical thickness of the plasma is $d = \int_0^t v_2 dt$. Based on the absorption law (5.55), the optical thickness is again found to be constant and equal to 1/4. The density and pressure of the plasma turn out to be

$$\rho_2 \simeq (\alpha d/B_3 t)^{3/8} I^{1/4} ; \quad p_2 \simeq (\alpha d/B_3 t)^{1/8} I^{3/4} \tag{5.59}$$

where $B_3 = B_2 [(1+Z)R/M]^{3/2}$. The conservation laws for mass and momentum at the shock front read (in laboratory coordinates)

$$\rho_0 u_S = \rho_1 (u_S - v_1) , \tag{5.60}$$

$$\rho_0 u_S{}^2 = \rho_1 (u_S - v_1)^2 + p_1 , \tag{5.61}$$

where u_S is the shock velocity and where the pressure p_0 of the undisturbed solid has been neglected. The pressure p_1 inside the shocked region is approximately equal to p_2, which, in view of its weak time dependence, can be assumed constant. Solving (5.60, 61) with $p_1 = p_2 \equiv p$ yields

$$v_1 = \sqrt{\left[1 - \frac{\rho_0}{\rho_1}\right] \frac{p}{\rho_0}} , \tag{5.62}$$

$$u_S = \sqrt{\frac{\rho_1}{\rho_1 - \rho_0} \frac{p}{\rho_0}} . \tag{5.63}$$

The surface of the solid retreats due to compression of the underlying material (at velocity v_1), as well as due to evaporation. The velocity due to evaporation alone is negligible and thus

$$u_v = \frac{v_2 \rho_2}{\rho_1} \simeq v_1 \sqrt{\frac{u_S \rho_0 \rho_2}{v_1 \rho_1{}^2}} \ll v_1 \tag{5.64}$$

where on the RHS use has been made of (5.62, 63) as well as the fact that $\rho_2 \ll \rho_0 \leq \rho_1$. The energy given to the solid consists essentially of the compression work, $\simeq p v_1$, which itself can be shown to be a small fraction (of the order of $(\rho_2/\rho_0)^{1/2} \ll 1$) of the pulse energy. Nevertheless, this small fraction is sufficient for ionizing the solid provided

$$p v_1 > \rho_0 u_S E_I N_A / M \quad \text{or} \quad I > (B_3 t)^{1/6} (\rho_0 E_I N_A / M)^{4/3} \ .$$

Caruso and *Gratton* estimated this to be satisfied for $I > 10^{11}$ W/cm^2 for a hydrogen isotope target, for which the model is expected to hold for fluxes between 10^9 and 10^{14} W/cm^2 in the case of ruby-laser light. As an illustration of this regime, Fig.5.16 shows density, velocity and temperature profiles from a numerical simulation by *Mulser* [5.66] (based on a more detailed model) for a hydrogen target shaped as a thin foil, irradiated by a ruby-laser pulse. The scaling at shorter wavelengths was discussed by *Ng* et al. [5.67]. Propagation of shock waves through solid material and their capability to produce spalling upon reflection at the back surface of irradiated slabs has also been investigated [5.68]. *Gupta* [5.69] used laser-induced spallation to measure quantitatively the adhesive strength of a planar interface between a substrate and its coating.

If the irradiance is further increased the pressure and density, and hence the temperature of the plasma increase, according to (5.59). The associated reduction in the absorption coefficient is balanced by an increase of the physical length of the plasma only as long as the plasma motion remains one dimensional, i.e., as long as d is smaller than about one focal-spot diameter. At the same time, the plasma density cannot increase beyond the cutoff density, at which the plasma turns opaque to the laser beam. The cutoff density is

$$\rho^* = \frac{M m_e \epsilon_0 \omega^2}{N_A Z e^2} \tag{5.65}$$

where ω is the laser frequency. Since the plasma temperature increases with its density, the maximum achievable temperature is reached when $\rho_2 = \rho^*$.

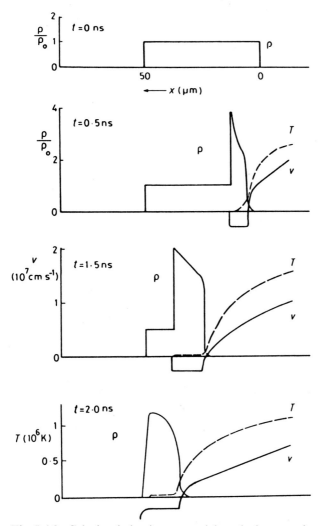

Fig.5.16. Calculated density ρ, particle velocity v and temperature T in a solid hydrogen foil irradiated by a ruby-laser pulse of 10^{12} W/cm^2 (incident from the right). The shock wave velocity is $2.7 \cdot 10^6$ cm/s [5.65]

From (5.58, 59) the maximum temperature is found to scale with the 2/3 power of the incident flux

$$RT_{2(\max)} = \frac{M}{1 + Z}\left[\frac{I}{\rho^*}\right]^{2/3} . \tag{5.66}$$

155

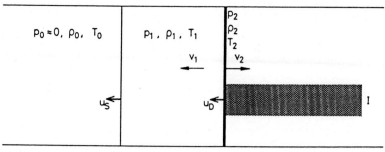

Fig.5.17. Target structure in the deflagration regime. Subscripts 0, 1, and 2 have the same meaning as in Fig.5.15. u_D is the velocity of the deflagration front, u_S that of the shock front (Laboratory coordinates)

Once the plasma is hot enough to be transparent to the laser pulse, volume absorption in the self-regulating plasma is replaced by sheet absorption at the interface between regions 1 and 2 in Fig.5.15, where the density equals the cutoff density. The resulting structure can be described as a laser driven deflagration wave.

5.4.2 Laser-Driven Deflagration Wave

A *deflagration wave*, like a detonation wave, is a discontinuity, but it is a rarefraction wave which moves at subsonic speed [5.70]. It is always preceded by an ordinary shock wave. The situation can be described by two sets of conservation laws for the two interfaces separating the regions 0, 1, and 2 (Fig.5.17). We have for the shock wave (neglecting p_0)

$$\rho_0 u_S = \rho_1 (u_S - v_1) , \tag{5.67}$$

$$\rho_0 u_S^2 = p_1 + \rho_1 (u_1 - v_1)^2 , \tag{5.68}$$

$$H_0/M + u_S^2/2 = H_1/M + (u_S - v_1)^2/2 . \tag{5.69}$$

For the deflagration wave, where the flux I is absorbed, we have

$$\rho_1 (u_D - v_1) = \rho_2 (u_D + v_2) , \tag{5.70}$$

$$p_1 + \rho_1 (u_D - v_1) = p_2 + \rho_2 (u_D + v_2)^2 , \tag{5.71}$$

$$H_1/M + (u_D - v_1)^2/2 = H_2/M + (u_D + v_2)^2 - I/\rho_2 (u_D + v_2) . \tag{5.72}$$

Here ρ_2 is taken to equal the cutoff density ρ^*. Assuming further that the compression by the shock wave is at its maximum possible value, $\rho_1/\rho_0 = (\gamma+1)/(\gamma-1)$ [5.9], neglecting the ionization energy and taking v_2 equal to the local sound velocity, we can express the velocities of the shock wave and the deflagration wave in terms of the irradiance [5.14]

$$u_S = \sqrt{\frac{(\gamma + 1)(\gamma' - 1)}{\gamma'} \frac{\rho^*}{\rho_0} \left[\frac{\gamma' + 1}{\gamma' - 1} \frac{I^2}{2\rho^{*2}} \right]^{1/6}}$$

$$(5.73)$$

$$u_D = u_S \frac{2}{\gamma + 1}$$

where γ and γ' denote the effective adiabatic coefficients of the solid and the plasma, respectively. If, as an example, a Nd-laser pulse of $I = 10^{14}$ W/cm^2 is incident on a deuterium target ($\gamma = 7/5$, $\gamma' = 5/3$) we have $\rho^* = 3.4 \cdot 10^3$ g/cm^3, and the model gives $u_S \simeq 10^7$ cm/s, $u_D = 0.83 \, u_S$. The ion temperature in the plasma is again found to scale with $I^{2/3}$ and reaches $kT_2 \simeq 3.5$ keV in the example.

5.4.3 Inertial Confinement

Let us now consider the situation for pulses in the ps regime where strong heating of the dense material occurs before hydrodynamic expansion has even started. The plasmas produced in this regime have essentially the same density as the solid.

Suppose a powerful pulse with a zero rise-time is switched on at t = 0 and pumps energy into the electrons of the material. The electron-ion collision time in a fully ionized plasma varies like $T_e^{2/3}$ and is of the order of 10^{-13} s for keV electrons. The electron-ion energy relaxation time is larger than this by a factor of the order of the ion mass divided by the electron mass, and hence it is likely to be longer than a mode-locking pulse. The ions will thus stay essentially cold at first and the only path for energy dissipation is heat conduction into the bulk material. Heat conduction is as fast as thermalization, while hydrodynamic motion is much slower – a characteristic time for the latter is the time it takes a sound wave to traverse the heated layer.

The electronic heat conductivity in a plasma is strongly temperature dependent, $K \propto T_e^{5/2}$ [5.71]. The resulting temperature distribution is a "heat wave", characterized by a sharply rising leading edge moving into the solid, followed by a plateau where the temperature is high and nearly uniform. The coordinate z_H of the leading edge of the heat wave at time t after the pulse is switched on, is given by [5.72]

$$z_H \simeq 9 \cdot 10^{10} I^{5/9} (t/N_e)^{7/9} \text{ cm} \quad \text{[cgs]} \tag{5.74}$$

where I is measured in erg/cm^{2}s and N_e in cm^{-3}. The average electron temperature behind the heat wave is

$$T_e \simeq 6 \cdot 10^4 I^{4/9} (t/N_e)^{2/9} \text{ K} \quad \text{[cgs]} . \tag{5.75}$$

While the heat wave develops, energy is fed to the ions and the heated layer begins to expand. The time during which the heat-wave regime lasts can be estimated as the time it takes the rarefraction wave, moving at the sound velocity, to catch up with the heat wave. This happens after a time

$$t_0 \simeq 2 \cdot 10^{13} I A^{3/2} N_e^{-2} \text{ s} \quad \text{[cgs]} \tag{5.76}$$

where A is the atomic number. Setting $A = 1$ and $N_e = 5 \cdot 10^{22}$ cm^{-3} gives, e.g., for

I [W/cm^2]	10^{10}	10^{12}	10^{14}	10^{16}
t_0 [s]	10^{-15}	10^{-13}	10^{-11}	10^{-9}

The above formulas hold for constant irradiance; for pulses much shorter than t_0 a somewhat different scaling is found [5.73].

We have so far neglected the problem of absorption of that light which reaches the region of the cutoff density. For a plasma at or near solid-state density almost total reflection of the light should be expected. Instead, strong absorption is observed experimentally. This is thought to be due to nonlinear optical effects in which electromagnetic and electrostatic plasma waves are excited. These, in turn, cause turbulence in the motion of the electrons and lead to far more absorption than expected from the normal electron-ion collisions. The effect has been described in terms of a phenomenological effective collision time [5.74]

$$1/\tau = 1/\tau_e + 1/\tau_p \tag{5.77}$$

where τ_e is the normal electron-ion collision time and τ_p characterizes the damping of the induced plasma waves. Particularly efficient coupling to longitudinal electrostatic plasma oscillations is observed if the beam is incident obliquely to the density gradient created by the hydrodynamic motion, since there is then a nonzero component of the electric field vector in the direction of the plasma oscillation. The effect is known as resonance absorption because there is an optimum incidence angle [5.75]. Damping of the electrostatic waves tends to produce "hot" (nonthermal) electrons and ions which decouple from the heat wave and limit the degree of inertial confinement achievable. Detailed discussions on these topics can be found in the literature [5.76].

5.5 Pulsed Laser Deposition

Having discussed the formation and evolution of laser plasmas in some detail, we shall now return to the solid state for this final section, closing full circle our parade of laser-induced phase transitions.

The basic arrangement in Pulsed Laser Deposition (PLD) is depicted in Fig.5.18. Some target material – usually placed inside a chamber with a controlled atmosphere – is irradiated obliquely by a focussed high-power laser pulse. A vapor plume forms and expands perpendicularly away from the target. Several cm from the impact spot a substrate is placed that intersects the expanding plume. Some of the vapor condenses and forms a deposit on the substrate. It is this deposit that we shall now consider.

PLD is basically just a physical vapor deposition scheme in which a laser happens to furnish the energy. Although the first attempts at PLD were reported as early as 1965 – as soon as high-power ruby lasers became available – the technique has attracted widespread interest only since the late eighties, mainly due to spectacular successes with high-T_c superconductor materials. The list of materials deposited today by PLD includes such diverse items as optical coatings, compound semiconductors, oxide superconductors and bioceramics. Before discussing the PLD process itself, let us briefly look into the basics of film growth from a vapor.

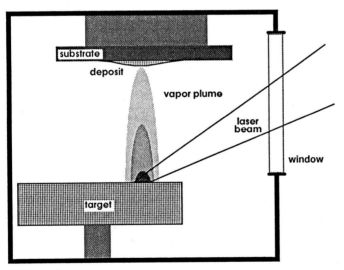

Fig.5.18. Experimental arrangement in pulsed laser deposition

5.5.1 Solid Film Growth from a Vapor

Vapor condensation is a nucleation-and-growth phenomenon like crystallization from a melt (Sect.4.1) or evaporation (Sect.5.1). Neglecting the fleeting liquid that may form from the vapor first, the overall driving force for condensation of the solids is the free energy difference between solid and vapor, $\Delta G_{sv} = G_s - G_v$. By means of the Clausius-Clapeyron law this can be expressed in terms of the pressure p or the molar flux j, see (5.4,6):

$$\Delta G_{sv} = -RT \ln[p/\bar{p}(T)] = -RT \ln[j/\bar{j}(T)] \tag{5.78}$$

In the present context of film condensation from a vapor, p is to be interpreted as the actual (local) vapor pressure, T as the substrate temperature, $\bar{p}$ as the equilibrium vapor pressure at temperature T, while j is the actual (molar) condensation rate and $\bar{j}$ the reevaporation rate, equal to the equilibrium condensation rate. We are, of course, interested in the regime where the vapor is undercooled or supersaturated, $(p > \bar{p})$ and hence $\Delta G_{sv} < 0$. In this regime the quantity $-\Delta G_{sv}$ is referred to as the *supersaturation*.

The first step of condensation is nucleation. The theory of crystal nucleation from a vapor is quite similar to the one for nucleation from a melt, as described in Sect.4.1.5. Here, as there, the nucleation rate depends upon the size of a "critical nucleus" which is determined by a balance between volume and surface forces (4.14). The difference to nucleation from a melt is that now the interfacial energies are much larger. Hence, homogeneous nucleation is more difficult (in fact, almost negligible) and "wetting" effects lowering the nucleation barrier are crucial. In the present context we are thus interested in heterogeneously nucleated condensation with the substrate as the seed.

To describe heterogeneous nucleation, the single interfacial energy σ used in connection with homogenous nucleation from a melt (Sect.4.1.5) needs to be replaced by the three interfacial energies σ_{sm}, σ_{sv} and σ_{mv} between the three phases present – the substrate (subscript s), the condensing material (m) and the vapor (v). The minimum-energy shape of a nucleus is no longer a sphere but a "cap" – the flatter the smaller the contact-angle α given by (Fig.5.19)

$$\cos(\alpha) = \frac{\sigma_{sv} - \sigma_{sm}}{\sigma_{mv}}. \tag{5.79}$$

The size of the critical nucleus, for reasons discussed in Chap.4, depends on the driving force, and therefore, by virtue of (5.78), on the deposition rate as well as on the substrate temperature. The relatively large nuclei characteristic of small supersaturation create isolated patches (islands) of film on the substrate which subsequently grow together. As the supersaturation

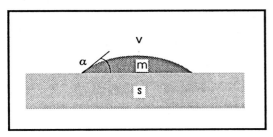

Fig.5.19. Cap-shaped nucleus showing contact angle α. (s: substrate, v: vapor, m: deposit)

increases, the critical nucleus shrinks until its height reaches one atomic diameter[3] and its shape is that of a two-dimensional layer. The transition from an island-type, or "3-D", to a layer-by-layer, or "2-D", type of nucleation is favored by good wetting between film and substrate (small α), as realized with atomically clean substrates chemically identical with the deposit. Only for very large supersaturation is the layer-by-layer nucleation regime also accessible with incompletely wetted foreign substrates [5.78].

Apart from nucleation, film formation also involves growth, mainly lateral growth. Since arriving vapor atoms (called "adatoms") must, on average, diffuse several atomic distances before sticking to a stable position within the newly forming film, lateral growth depends on a sufficient surface mobility of the adatoms. Surface diffusion is a thermally activated process determined by the substrate temperature. High temperature favors rapid and defect-free crystal growth, whereas at low temperature or large supersaturation crystal growth may be overwhelmed by impingement, resulting in disordered or even amorphous structures.

A useful, if necessarily simplified, framework to characterize the influence of the two main parameters substrate temperature T and impingement flux j on film deposition is due to *Kaschiev* [5.78]. He defined a quantity termed N_{99} as *the average number of monolayers required for the growing film to reach continuity* (actually defined as 99% of coverage). The value of N_{99} reflects both nucleation and growth rates, and Fig. 5.20 shows its dependence on the substrate temperature and impingment flux. Theory predicts that below a "critical" value near $N_{99} = 2$ film formation occurs layer by layer, whereas above it involves formation and coalescence of progressively thicker islands. One would thus choose the former regime to produce continuous monolayers and the latter to achieve single-crystalline film growth.

---

[3] At very large supersaturation the critical nucleus predicted by classical theory may turn out to be smaller than one atom. In this situation atomistic models which use atomic potential energies instead of bulk free energies are better suited [5.77]. Conceptually, however, the conclusions are the same.

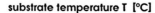

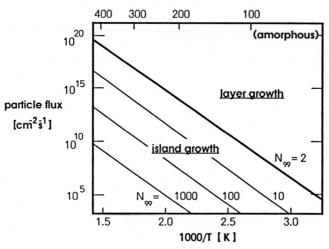

Fig.5.20. Growth diagram illustrating regimes of layer-by-layer growth and island growth depending on substrate temperature T and impingement particle flux [5.78, 79]

5.5.2 The PLD Process

In order to apply the above considerations quantitatively to a PLD experiment, one would have to know the relevant parameters of the vapor which define the quantities p and j in (5.78). This, of course, is no trivial matter. As we have discussed in most of this chapter, parameters like particle density and kinetic energy, degree of ionisation, temperature and pressure in an evolving laser plume tend to vary widely depending on the absorbed laser fluence, pulse duration and wavelength. Moreover, the vapor usually consists of several kinds of particles normally not in thermodynamic equilibrium.

No quantitative general theory relating all vapor properties of interest to the laser pulse and material parameters is available. The best we can do (as so often) is to pull together what is known from theory and experiment in order to obtain at least a qualitative picture of the complex phenomena leading to PLD.

Whatever the details of the interaction, the laser impact creates a high-pressure vapor plume. The plume, somewhat similar to a rocket exhaust, expands away from the impact zone with a strongly forward directed velocity distribution, often approximated by a distribution of the form $\cos^n(\theta)$, θ being the angle to the target normal and n a number between about 5 and 12 [5.80]. While the laser pulse lasts, it interacts with the plume in various ways depending on irradiance and wavelength, as discussed in previous sections. As a result, the plume may heat up far beyond the surface temperature. Single- or multiple-charged and energetic ions may be produced and shock waves may develop.

After the laser pulse the vapor plume expands further, cools, and eventually dissipates. Somewhere in this sequence the plume front meets the substrate which precipitates condensation. In order to determine the vapor properties at this point, the plume is usually treated as an adiabatically expanding ideal gas (Sect. 5.2.2). Simplified calculations of plume expansion and resulting film deposition parameters have been presented [5.81, 82]. In their analytical treatment, *Anisimov* et al. [5.81] assumed a self-similar adiabatic expansion mode and predicted a film profile of the form

$$h(r) = \frac{M_0 k^2}{2\rho rz^2} \left[1 + k^2 \frac{r^2}{z^2} \right]^{-3/2} \tag{5.80}$$

Here M_0 is the total initial mass of the plume (a function of the laser pulse energy), ρ is the film solid density, z is the distance between the target and the substrate (Fig. 5.18) and k is a (tabulated) function of the adiabatic index of the vapor, the focal spot radius and the pulse duration, roughly with values between 2 and 10. For typical parameters (50 ns laser pulse, one to several J/cm^2 of absorbed fluence, $z = 5$ cm, $M_0 = 1...10 \mu g$, $\rho = 2.5$ g/cm^2, $k = 2$) the predicted film thickness deposited per pulse is $1...10$ Å thick at the center with a radius at half-thickness of 2 cm. The particle flux j/N_A in this process ranges between 10^{26} and 20^{22} $cm^{-2} s^{-1}$ and the supersaturation ΔG_{sv} reaches some 10^5 J/mol. By suitably dosing the pulse energy, as this estimate indicates, individual monolayers can readily be deposited.

These are impressive numbers. Indeed, the plumes created by short laser pulses differ significantly from the near-equilibrium vapors present in conventional (crucible-type or electron-gun driven) evaporation systems. Firstly, PLD vapors are hotter – the surface reaches temperatures well in excess of the equilibrium evaporation temperature of the target material. As a consequence, the vapors tend to be fully dissociated even for molecular targets [5.83]. More important, however, is that their elemental composition is equal to that of the target – no matter what the vapor-pressure curves of the constituents (for the explanation of this fact see the end of Sect. 5.1.1). This feature, which is largely responsible for the popularity of PLD among materials scientists, enables them to deposit stoichiometric compound semiconductors like HgCdTe – which consist of species with vapor pressures differing by several orders of magnitude – simply from a target of the same material.

Another property of PLD vapors, as opposed to equilibrium ones, is the presence of fast (sometimes falsely called *superthermal*) atoms, as well as excited particles including ions. The high atom velocities are a natural consequence of the high gas pressures reached and are usually found to be Maxwellian distributed around stream velocities of some tens of km/s, in accordance with gas-dynamic considerations. Deviations from Maxwellian behavior is only found for UV laser pulses, indicative that here evaporation is partly due to photolytic desorption [5.84].

Ion velocities are even higher, apparently due to acceleration in electric fields building up in the plume. The fraction and energy of energetic particles increases with the irradiance. Typical ion fractions for UV lasers are less than 5% for irradiances below 100 MW/cm^2 but approach 100% at 10 GW/cm^2; infrared lasers reach full ionisation even earlier. Ion kinetic energies in the range of $10 \div 1000$ eV, one to two orders of magnitude higher than for neutrals, seem to be typical [5.82]. There are, of course, also large numbers of free electrons present in the ionized vapor, but little is known about their properties and impact on film formation. As to the ions, their influence on film formation is mainly beneficial – they seem to provide extra energy to the growing surface, enhancing surface mobility and hence improving crystal quality and film density even on cold substrates. High-energy ions, on the other hand, have been blamed for damaging the growing film by defect formation and sputtering.

Yet another species populating PLD plumes is unconditionally unpopular – melt droplets. Tiny droplets may arise due to condensation in the supersaturated vapor (Sect.5.2.2) [5.85], but the main contribution seems to be melt ejected from the target surface, either by boiling or by evaporation recoil (Sect.5.2.3). Droplets are detrimental to the quality of the film and unsuitable for applications. The fraction of droplets in the plume can be reduced by appropriately selecting the fluence [5.86], but this may not always be practical. Fortunately, droplets move at much lower speeds than gas particles and can be filtered out of the plume by means of mechanical velocity filters [5.87]. Alternatively, the use of a second laser beam to evaporate droplets has been proposed [5.88].

Otherwise, PLD is experimentally quite straightforward. Apart from the laser, a simple target chamber with an optical window and fixtures for the substrate (perhaps heatable) and the target (preferentially rotatable to expose a fresh spot for every laser shot) is all that is required. A good vacuum is not necessary – at deposition rates of $10^4 \div 10^6$ Å/s, impurity incorporation is always negligible compared to deposition. PLD also works well under relatively high pressures ($1 \div 100$ mTorr) of reactive gases.

What materials can be deposited by PLD? As it appears, in principle almost any – for detailed reviews see [5.83, 88]. Early reports emphasized optical coatings based on oxides, chalcogenides and metals, which were lauded for being denser and smoother than thermally evaporated ones [5.88]. The inherent monolayer resolution of PLD – supported by advances in laser technology – was soon put to use in the fabrication of layered structures like tailored bandgap (superlattice) semiconductors [5.90], X-ray mirrors [5.91] or Bi-CdTe pairs with barrier layers as thin as 20 Å [5.91], just to mention a few examples. However, PLD has turned out to be at its best in the deposition of complex multicomponent ceramic crystals, in which stoichiometry is crucial but difficult to achieve by standard deposition techniques. The most notable exampls are the oxide ceramic superconductors like $YBa_2Cu_3O_7$ (YBCO) and others, but also ferroelectrics like $BaTiO_3$ or even $(Pb, La)(Zr, Ti)O_3$, ferrites and many others [5.89]. The laser of choice in this work was the excimer, the radiation of which is efficiently absorbed in the ceramic target.

Singh and *Narayan* [5.82] investigated PLD of YBCO materials by ns-excimer laser pulses in detail. Plasma temperatures of 10 000 K and particle velocities of several km/s (scaling approximately as the cube root of laser pulse energy) are characteristic of this regime. Because lighter atoms move faster than heavier ones (although the dependence is weaker than the inverse-square-root mass dependence expected for non-interacting particles), slight spatial variations of film composition tend to build up, with lighter species enriched at the center of the deposit, even though the evaporation itself is stoichiometric. The effect becomes smaller at larger target-substrate distances and higher background pressures since collisions tend to equalize species velocities.

The stoichiometric evaporation achieved by PLD thus does not automatically guarantee exactly stoichiometric films in all cases. Deviations may also occur due to differing elemental sticking coefficients, reevaporation of volatile elements or preferential sputtering by fast ions. In order to enhance incorporation of oxygen and stabilize the desired tetragonal YBCO phase between laser pulses, oxygen background gas or even oxygen jets directed at the substrate have been employed [5.88, 93]. Apart from oxides, a similar kind of "reactive PLD" has been demonstrated for nitride formation in a N_2 atmosphere [5.94]. Yet another approach to promote incorporation of a volatile element is the use of a second target rich in that element [5.95].

The list of experimental variants of the PLD method as well as the range of materials successfully deposited is certain to grow in the future. The technique is in its infancy, at least as far as its quantitative physical understanding is concerned. There remains plenty to be done.

A. Appendix

A.1 Selected Material Data

In the following we provide a short compilation of optical, thermophysical and thermodynamic material properties often used in treating light-material interaction phenomena. The data are meant as a quick reference for the convenience of the reader. Original sources should be consulted when accurate data for quantitive calculations are needed.

Table A.1. Optical absorption lengths ($1/\alpha$) and reflectances (R) of semiconductors (c: crystalline, a: amorphous, l: liquid), insulators and evaporated metal films at room temperature for various wavelengths. Data from [A.1] and other sources. Semiconductor data depend on purity and the method of preparation, and are meant as typical values only

Wavelength:	$0.25\,\mu m$		$0.5\,\mu m$		$1.06\,\mu m$		$10.6\,\mu m$	
	$1/\alpha$	R	$1/\alpha$	R	$1/\alpha$	R	$1/\alpha$	R
GaAs (c)	6nm	0.6	100nm	0.39	$70\,\mu m$	0.31	>1cm	0.28
Ge (c)	7nm	0.42	15nm	0.49	$200\,\mu m$	0.38	>1cm	0.36
Ge (a)	10nm	0.48	50nm	0.47	$1\,\mu m$	0.42	>1cm	0.34
Si (c)	6nm	0.61	500nm	0.36	$200\,\mu m$	0.33	1mm	0.30
Si (a)	10nm	0.75	100nm	0.48	$1\,\mu m$	0.35	>1cm	0.32
Si (l)			8nm	0.72	13nm	0.72		
KCl	>1cm	0.05	>1cm	0.04	>1cm	0.04	>1cm	0.03
SiO_2	>1cm	0.06	>1cm	0.04	>1cm	0.04	$40\,\mu m$	0.2
Ag	20nm	0.30	14nm	0.98	12nm	0.99	12nm	0.99
Al	8nm	0.92	7nm	0.92	10nm	0.94	12nm	0.98
Au	18nm	0.33	22nm	0.48	13nm	0.98	14nm	0.98
Cu		(0.1)	14nm	0.62	13nm	0.98	13nm	0.99
Ni		(0.15)	12nm	0.62	15nm	0.67	37nm	0.97
W	7nm	0.51	13nm	0.49	23nm	0.58	20nm	0.98

Table A.2. Optical breakdown threshold irradiance I_B for various wavelengths (λ) and pulse durations (t_p) in insulating and semiconducting materials (B: bulk, S: surface). Data selected from a review article by *Smith* [A.2] where references to the original literature can be found

Substance		λ [μm]	t_p		I_B [W/cm^2]	
Al$_2$O$_3$	B	1.06	4.7	ns	1.8	$\times 10^{11}$
	B	1.06	30	ps	7.7	$\times 10^{11}$
	S	1.06	4.7	ns	4.6	$\times 10^{10}$
Diamond	B	10.6	1	s	1	$\times 10^6$
Silica	B	1.06	4.7	ns	1.1	$\times 10^{11}$
	B	1.06	30	ps	1.3	$\times 10^{12}$
	B	0.53	21	ps	1.4	$\times 10^{12}$
	B	0.355	20	ps	2$\div$9	$\times 10^{12}$
	S	1.06	125	ps	8	$\times 10^{10}$
glass:BK-7	S	1.06	125	ps	7.4	$\times 10^{10}$
glass:BSC-2	B	1.06	125	ps	7.4	$\times 10^{10}$
	S	1.06	4.7	ns	5$\div$8.8	$\times 10^{10}$
glass:Pyrex	S	1.06	125	ps	6.4	$\times 10^{10}$
PMMA plastic	S	1.06	12	ns	1.6	$\times 10^{10}$
NaCl	B	10.6	600	ns	0.46$\div$1.6	$\times 10^{10}$
	B	10.6	80	ns	1.55	$\times 10^{10}$
	B	1.06	4.7	ns	2.2	$\times 10^{10}$
	B	1.06	30	ps	9	$\times 10^{10}$
	B	1.06	15	ps	6.3	$\times 10^{11}$
	B	0.53	21	ps	6.3	$\times 10^{11}$
ZnSe	B	10.6	92	ns	4$\div$5	$\times 10^8$
	B	1.06	4.7	ns	1.2	$\times 10^9$
	S	10.6	92	ns	3.5	$\times 10^8$
H$_2$O (liq)	B	1.06	30	ps	1.6	$\times 10^{12}$
D$_2$O (liq)	B	1.06	30	ps	8	$\times 10^{11}$
GaSa	S	10.6	0.2	s	1.3	$\times 10^3$
GaP	S	10.6	0.2	s	3.6	$\times 10^3$
Ge	S	10.6	0.2	s	8.5	$\times 10^3$
	S	10.6	1.4	ns	3	$\times 10^9$
Si	S	10.6	0.2	s	1$\div$2	$\times 10^4$

Table A.3. Melting points and absorptances [%] of some elements at various temperatures. $(1-R)^{(fe)}$ values calculated from (2.46), $(1-R)^{(exp)}$ from literature data. Based on data from [A.1, 6-8], and other sources

Metal	T_{sl}	$(1-R)^{fe}$				$(1-R)^{exp}$	
	[K]	300K	T_{sl}(sol)	T_{sl}(liq)	$T_{sl}+500K$	$\lambda = 1\,\mu m$, 300K	
						(lowest)	(practical)
Ag	1234	0.4	2.0	4.3	5.4	0.6	1
Al	933	1.1	4.4	10.0	13	5.6	6
Au	1336	0.6	3.5	8.3	10.3	1.4	2
Cu	1356	0.5	2.8	5.8	7.3	1.0	2
Sn(gray)	505	0.1	0.2	0.4	0.5	---	54
W	3655	0.8	17.1	18.5	---	42	45
Zn	693	2.1	6.0	13.5	13.4	---	43

A.2 Green's Functions for Solving the Heat-Flow Equation

The starting point of our analysis is the Green's function for a point-like heat source. For the boundary conditions (3.5) it is given by [A.9]

$$g_0 = \frac{V}{\pi^{3/2} c_p \beta^3} \exp\left[-\frac{(x-x')^2 + (y-y')^2}{\beta^2} \right]$$

$$\times \sum_{n=-\infty}^{\infty} \left[\exp\left(-\frac{(2nL-z-z')^2}{\beta^2} \right) + \exp\left(-\frac{(2nL-z+z')^2}{\beta^2} \right) \right] \quad \text{(A2.1)}$$

with $\beta = 2(\kappa|t-t'|)^{1/2}$. From (A2.1) Green's functions for any desired source geometry can be constructed. We are interested in a distributed heat source with an energy density decreasing exponentially inside the material. A particularly simple case is that of a vanishing absorption length $\alpha^{-1} = 0$. We refer to this geometry, which is described by (A2.1) with $z' = 0$, as a *surface source*. For future reference we denote the sum in (A2.1) for $z' = 0$ as Υ_1

$$\Upsilon_1 = 2 \sum_{n=-\infty}^{\infty} \exp\left(\frac{(2nL-z)^2}{\beta^2} \right). \quad \text{(A2.2)}$$

Table A.4. Thermodynamic data of selected elements, from [A.1, 3, 4] and other sources (M: molar mass; N: number density; ρ_s, ρ_ℓ: mass densities of solid and liquid, respectively; $T_{s\ell}$: solid-liquid equilibrium temperature; $T_{\ell v}$: liquid-vapor equilibrium temperature; $\Delta H = \int c_p\, dT$; $\Delta H_{s\ell}, \Delta H_{\ell v}$: latent heats)

	M [g/mole]	N (300K) [10^{-22} cm^{-3}]	ρ_s (300K) [g/cm^3]	ρ_l (T_{sl}) [g/cm^3]	T_{sl} [K]	T_{lv} [K]	ΔH 300K→T_{sl} [kJ/mole]	ΔH_{sl} [kJ/mole]	ΔH T_{sl}→T_{lv} [kJ/mole]	ΔH_{LV} [kJ/mole]
Ag	107.9	5.85	10.50	9.30	1234	2485	23.3	12.0	37.2	254.2
Al	27.0	6.02	2.70	2.37	933	2793	17.2	10.7	52.4	284.1
Au	197.0	5.91	19.3	17.3	1338	3130	29.2	12.6	52.6	310.6
Cd	112.4	4.60	8.6	8.0	594	1040	8.2	6.1	13.3	99.4
Co	58.9	8.99	8.9	7.7	1768	3201	54.0	15.9	49.8	389.6
Cr	52.0	8.33	7.1	6.5	2130	2945	62.5	14.7	32.1	305.4
Cu	63.5	8.48	8.96	8.0	1358	2836	29.5	13.0	46.4	304.8
Fe	55.8	8.48	7.87	7.0	1809	3135	65.0	19.4	55.5	353.8
Ge	72.6	8.43	5.32	7.0	1809	3107	25.8	34.8	58.0	284.7
Mg	24.3	4.30	1.74	1.57	922	1363	17.7	9.2	14.8	131.9
Ni	58.7	9.13	8.9	7.8	1726	3187	47.4	17.6	56.3	378.8
Pb	207.2	3.29	11.7	10.6	601	2023	8.5	4.8	40.4	178.0
Pd	106.4	6.77	12.0	10.7	1825	3237	46.4	17.2	53.2	372.6
Pt	195.1	6.62	21.5	18.9	2043	4100	53.5	21.9	77.5	510.8
Sb	121.8	3.27	6.68	6.5	904	1860	16.6	20.1	30.1	195.5
Si	28.1	4.98	2.33	2.53	1685	3514	36.0	50.3	136	385.6
Sn	118.7	3.70	7.30	6.96	505	2876	5.9	7.1	72.5	230.2
Ta	180.9	5.53	16.6	15.0	3287	5731	95.3	31.5	95.5	753.4
Ti	47.9	6.68	4.5	4.2	1943	3562	54.1	19.3	52.9	422.9
W	183.9	6.32	19.3	17.7	3680	5628	102.6	35.3	77.2	774.6
Zn	65.4	6.55	7.14	6.66	693	1180	10.8	6.7	15.3	114.8

Table A.5 Selected binary systems grouped according to the types showing similar alloying behavior

---

(i) Extended or complete solid-solution formers

Ag-Au	Ag-Pd	Au-Cu	Au-Pd	Co-Fe	Co-Mn	Co-Ni	Co-Pd	Co-Pt
Cr-Fe	Cr-Ni	Cr-Pt	Cr-V	Cu-Ni	Cu-Pd	Cu-Pt	Fe-Mn	Fe-Ni
Fe-Pd	Fe-Pt	Fe-V	Ge-Si	Mn-Ni	Mn-Pt	Mn-V	Ni-Pd	Ni-Pt
Ti-W	V-W							

(ii) Compound formers

Ag-Al	Ag-Pt	Ag-Mg	Ag-Sn	Ag-Ti	Ag-Zn	Al-Au	Al-Co	Al-Cr
Al-Cu	Al-Fe	Al-Mg	Al-Mn	Al-Ni	Al-Pd	Al-Pt	Al-Ti	Al-W
Al-Zn	Au-Cr	Au-Fe	Au-Mg	Au-Mn	Au-Pb	Au-Sn	Au-Ti	Au-V
Au-Zn	Co-Cr	Co-Cu	Co-Mg	Co-Si	Co-Ti	Co-V	Co-W	Co-Zn
Cr-Ge	Cr-Mg	Cr-Mn	Cr-Pd	Cr-Si	Cr-Zn	Cu-Fe	Cu-Ge	Cu-Mg
Cu-Si	Cu-Sn	Cu-Ti	Cu-Zn	Fe-Ge	Fe-Si	Fe-Ti	Fe-W	Fe-Zn
Ge-Mg	Ge-Mn	Ge-Ni	Ge-Pd	Ge-Pt	Ge-Ti	Ge-V	Mg-Ni	Mg-Pb
Mg-Pd	Mg-Pt	Mg-Si	Mg-Sn	Mg-Zn	Mn-Pd	Mn-Si	Mn-Sn	Mn-Ti
Mn-Zn	Ni-Si	Ni-Sn	Ni-Ti	Ni-V	Ni-W	Ni-Zn	Pb-Pd	Pb-Pt
Pb-Ti	Pb-V	Pd-Si	Pd-Sn	Pd-Ti	Pd-Zn	Pt-Si	Pt-Sn	Pt-Ti
Pt-V	Pt-Zn	Si-Ti	Si-V	Si-W	Si-Zr	Sn-Ti	Sn-V	Sn-Zr
Ti-Zn	V-Zn	Zn-Zr						

(iii) Immiscible solids (simple eutectics)

Ag-Cu	Ag-Ge	Ag-Pd	Ag-Si	Al-Be	Al-Ge	Al-Si	Al-Sn	Au-Co
Au-Ge	Au-Si	Fe-Mg	Ge-Pb	Ge-Sn	Ge-Zn	Mg-Ti	Pd-Sn	Pd-V
Pd-W	Pt-W	Si-Sn	Si-Zn	Sn-Zn				

(iv) Immiscible melts

Ag-Cr	Ag-Fe	Ag-Ni	Ag-V	Ag-W	Al-Cd	Al-Pd	Au-Ru	Au-W
Co-Pd	Cr-Pb	Cr-Sn	Cu-Mo	Cu-Pb	Cu-V	Cu-W	Fe-Pb	Fe-Sn
Mg-V	Mg-W	Mn-Pb	Ni-Pb	Pb-Si	Pb-Zn	Sn-W	Zn-W	

Table A.6. Vapor pressure of selected elements at various temperatures. Data from [A.3, 5] and other sources

| | Vapor pressure [atm] at temperature [°C] | | | | | |
	1000	1500	2000	2500	3000	3500
Ag		0.01	0.3	3.5	18	70
Al			0.7	1.2	8	38
Au				0.12	1.3	7
Cd	0.02	1	9	35		
Cr			0.05	1.1	10	
Cu			0.03	0.6	5	25
Fe				0.3	3	18
Ge				0.15	2.5	12
Mg	0.4	15				
Ni			0.01	0.3	3.5	20
Pb		0.2	3.2	20		
Pt					0.02	0.25
Si				0.03	0.4	2.5
Sn			0.03	0.5	2.5	10
Ti					0.01	0.3	3
Zn	2.5	70				

In a material with a finite absorption length, light absorption represents a *penetrating source*, for which the appropriate Green's function is

$$g_1 = \alpha \int_0^L g_0\, e^{-\alpha z'} dz'$$

$$= \frac{\alpha V}{2\pi\, c_p\, \beta^2} \exp\left[-\frac{(x-x')^2 + (y-y')^2}{\beta^2} - \frac{\alpha^2 \beta^2}{4} \right] \Upsilon_2 \qquad (A2.3)$$

with

$$\Upsilon_2 = \sum_{n=-\infty}^{\infty} \exp[\alpha(2nL - z)]$$

$$\times \left[\operatorname{erfc}\left(\frac{(2n+1)L - z}{\beta} + \frac{\alpha\beta}{2} \right) - \operatorname{erfc}\left(\frac{2nL - z}{\beta} + \frac{\alpha\beta}{2} \right) \right]$$

$$+ \exp[-\alpha(2nL - z)] \qquad (A2.4)$$

$$\times \left[\mathrm{erfc}\left(- \frac{(2n+1)L - z}{\beta} + \frac{\alpha\beta}{2}\right) - \mathrm{erfc}\left(- \frac{2nL - z}{\beta} + \frac{\alpha\beta}{2}\right) \right] .$$

Note that for the semi-infinite solid $L = \infty$, and thus only the terms $n = 0$ in the sums Υ_1 and Υ_2 need to be retained. To describe heating by an extended laser beam the Green's functions given above must be integrated over the lateral beam distribution. If $f(r)$ is the radial irradiance profile we have (with $r^2 = x^2 + y^2$)

$$g_2 = \left[2\pi \int_0^\infty f(r')\,dr' \right]^{-1} \int_0^\infty \int_0^{2\pi} g_i(r,r')\,f(r')\,r'\,dr'\,d\phi \qquad (A2.5)$$

where g_i is one of the expressions (A2.1 or 3). (Note that the distance between a field point and a source point in cylindrical coordinates is $[(x-x')^2 + (y-y')^2]^{1/2} = (r^2 + r'^2 - 2rr'\cos\phi)^{1/2}$). We only consider two cases here, the *uniform* and the *Gaussian* source. Other beam profiles can be treated by the same procedure.

The **uniform source**, $f(r) = \text{constant}$, is somewhat unphysical since it is not normalizable. We override the difficulty by introducing an infinite surface S_∞, with the understanding that for an evaluation of (3.7) P_a/S_∞ is replaced by the source irradiance. Thus, we obtain for the surface source

$$g_{us} = \frac{V}{c_p S_\infty \beta \sqrt{\pi}} \cdot \Upsilon_1 \qquad (A2.6)$$

and for the penetrating source

$$g_{up} = \frac{\alpha V}{2 c_p S_\infty} \exp[-(\alpha\beta/2)^2] \cdot \Upsilon_2 . \qquad (A2.7)$$

For the **Gaussian sources** we set $f(r) = \exp(-r^2/w^2)$ and get for the surface source

$$g_{gs} = \frac{V}{\pi^{3/2} c_p \beta(\beta^2 + w^2)} \exp\left(- \frac{r^2}{\beta^2 + w^2}\right) \cdot \Upsilon_1 \qquad (A2.8)$$

and for the penetrating source

$$g_{gp} = \frac{V}{2\pi c_p (\beta^2 + w^2)} \exp\left(- \frac{r^2}{\beta^2 + w^2} + \frac{\alpha^2\beta^2}{4}\right) \cdot \Upsilon_2 . \qquad (A2.9)$$

To obtain a temperature distribution, the appropriate Green's function, together with a function describing the temporal pulse envelope $P_a(t)$, is inserted into (3.7) and the integral is solved.

A.3 Numerical Solution of the Heat-Flow Equation

Equations (4.4,5), along with the appropriate initial and boundary conditions, completely specify our problem. In a sense, they already represent a computational scheme. For an actual computer evaluation, the differential equation (4.5) must merely be transformed into a finite-difference equation. Here, instead of continuous functions of space and time, such as $T(z,t)$ or $\Delta H(z,t)$, one has discrete functions defined at points on a lattice into which the irradiated solid is divided. Since we are considering one-dimensional heat flow only (a generalization to three dimensions does not add anything new), the lattice is one-dimensional and may be visualized as dividing the solid into parallel layers with uniform internal temperature and material properties. Lattice points are separated by the finite length Δz, and function values are evaluated only after discrete time intervals Δt. For convenience, we introduce the volumetric enthalpy $W \equiv \Delta H/V$ (V is taken as a constant). A finite-difference version of (4.4,5) can now be written as

$$W_n^{i+1} = W_n^i + (\Delta t/\Delta z^2)[K_{n-1,n}^i(T_{n-1}^i - T_n^i)$$
$$- K_{n,n+1}^i(T_n^i - T_{n+1}^i)] + J_n^i , \qquad (A3.1)$$

and

$$T_n^i = T(W_n^i) . \qquad (A3.2)$$

Here X_n^i denotes the value of the quantity X at the lattice site n at time i (i.e., i time increments Δt after the chosen time origin). Further, $K_{n,n+1}^i$ specifies the conductivity at time i between the lattice points n and $n+1$. Eq.(A3.2) is an inverted version of (4.4) in the form of a table or an analytic approximation.

The procedure stipulated by (A3.1,2) works as follows. An initial distribution W_n^0, corresponding to the desired initial temperature distribution T_n^0, is specified and inserted into (A3.1). If there is heat production, i.e., if J_n^0 is nonzero, then a new distribution W_n^i is calculated, with $W_n^1 > W_n^0$. Next, the enthalpy distribution W_n^1 is converted into a temperature distribution T_n^1 by means of (A3.2), and inserted back into (A3.1). The procedure is repeated as many times as desired, yielding a new instantaneous temperature profile T_n^i after each time step. The boundary condition for an insulated slab is enforced in the calculation by adding two virtual lattice points at $n = 0$ and $n = N+1$, where $n = 1$ and $n = N$ are the outmost lattice points of the slab and setting $W_0^i = W_1^i$ and $W_{N+1}^i = W_{Na}^i$. The interface is conveniently defined as that lattice site which has absorbed or liberated 50 % of the latent heat of melting (the interface temperature is specified by (A3.2)). The position of the interface is recorded after each time step. There is almost complete freedom to readjust the material parameters or the heat source characteristics between computing steps, in order to allow, e.g., for a temperature-dependent conductivity, for temporal variations of the irradiance, or for absorption effects.

The advantage of such a code (termed an *explicit code* by the experts) is that it is straightforward and allows maximum flexibility in the choice of the physical parameters. The price paid for these advantages is that the increments Δt and Δz must satisfy the stringent condition

$$\Delta t < 0.5\,\Delta z^2\,(c_p/VK)_{minimum} \tag{A3.3}$$

everywhere and at all times. This means, in practice, that rather small time steps must be chosen if a good spatial resolution is desired. The reason for this condition has nothing to do with rounding errors as sometimes stated, but lies rather in the fact that the finite-difference equation (A3.1) has solutions which do not satisfy the original differential equation (4.5). Only those solutions of (A3.1) which also satisfy (A3.3) remain always real and finite.

A.4 Units and Symbols

Systeme International (SI) units are used in formulas, except when noted otherwise. Several familiar non-SI units have been retained in quoting numerical data. The values of some non-SI units employed in the book, expressed in SI-units, are

1 eV	=	$1.602 \cdot 10^{-19}$ J
1 at	=	$9.806 \cdot 10^4$ Pa
1 g/cm^3	=	10^3 kg/m^3
1 W/cm^2	=	10^{-4} W/m^2

The main symbols and frequently used subscripts are listed in the following. Equilibrium quantities are, where necessary, distinguished from actual ones by an overbar.

A.4.1 Constants

c	vacuum velocity of light ($2.998 \cdot 10^8$ m/s)
e	electron charge ($1.602 \cdot 10^{-19}$ C)
$\hbar$	Planck's constant/2π ($1.054 \cdot 10^{-34}$ J·s = $6.583 \cdot 10^{-16}$ eV·s)
k	Boltzmann's constant [$1.380 \cdot 10^{-23}$ J/K $\simeq (1/11605)$ eV/K]
m_e	free electron rest mass ($9.108 \cdot 10^{-31}$ kg)
N_A	Avogadro's number ($6.023 \cdot 10^{26}$ mole^{-1})
R	Gas constant ($8.314 \cdot 10^3$ J/K)
ϵ_0	dielectric constant ($8.854 \cdot 10^{-12}$ C/Vm)
σ_{SB}	Stefan-Boltzmann constant ($5.667 \cdot 10^{-8}$ W/m^{2}K^4)

A.4.2 Variables

A	particle accommodation factor
a	molecular diameter or lattice constant [m]
c_p	specific heat at constant pressure [J/mole K]
c_v	specific heat at constant volume [J/mole K]
D	diffusivity [m^2/s]
D_{amb}	ambipolar diffusivity [m^2/s]
E	energy [J]
E_F	Fermi energy [J]
E_g	band-gap energy [J]
E	electrical field strength [V/m]
ϵ	emissivity
F	fluence [J/m^2]
G	Gibbs free energy (free enthalpy) [J/mole]
g	partial Gibbs free energy (free enthalpy) [J/mole]
g	Green's function
H	magnetic field strength [A/m]
H	enthalpy [J/mole]
h	partial enthalpy [J/mole]
I	irradiance [W/m^2]
J	power density [W/m^3]
j	molar flux [moles/m^{2}s]
K	thermal conductivity [W/m K]
k	distribution coefficient
L	slab thickness [m]
ℓ	mean free path [m]
M	molar mass [kg/mole]
m	particle mass [kg]
N	particle density [m^{-3}]
n	complex refractive index

n_1, n_2	real and imaginary parts of **n**
P	power [W]
p	pressure [Pa]
R	reflection coefficient
r	coordinate, radius [m]
S	entropy [J/mole K]
S_T	Soret coefficient [K^{-1}]
T	temperature [K]
t	time [s]
t_p	pulse duration [s]
u	velocity [m/s]
u_s	sound velocity [m/s]
V	molar volume [m^3/mole]
v	velocity [m/s]
w	beam radius [m]
X	molar/atomic fraction
x	coordinate [m]
Y	nucleation rate [$m^{-3}s^{-1}$]
y	coordinate [m]
Z	ion charge number
z	coordinate [m]
α	absorption coefficient [m^{-1}]
Γ	damping constant [s^{-1}]
γ	$= c_p/c_v$ adiabatic index
δ	diffusion length [m]
ϵ	complex dielectric function
ϵ_1, ϵ_2	real and imaginary parts of ϵ
ζ	relative coordinate along z-axis [m]
θ	normalized temperature
κ	thermal diffusivity [m^2/s]
λ	wavelength [m]
ν_j	jump frequency [s^{-1}]
ρ	mass density [kg/m^3]
Σ	cross section [m^2]
σ	surface free energy [J/m^2]
σ	electrical conductivity [A/Vm]
τ	time constant [s]
Φ	heat flux [W/m^2]
Ξ	volume fraction
ω	angular frequency [s^{-1}]
ω_p	plasma frequency [s^{-1}]
[...]	equals one if the bracketed condition holds, zero otherwise

A.4.3 Subscripts

a	absorbed
cr	critical
d	dense phase (solid or liquid)
e	electron
h	hole
l	liquid
r	recombination
s	solid
v	vapor

References

Chapter 1

1.1 For recent surveys see, e.g.,
Laser Materials Processing – Industrial and Micrelectronics Applications, ed.
by E. Beyer, SPIE Proc. **2207** (1994)
Laser Assisted Fabrication of Thin Films and Microstructures, ed. by I. Boyd,
SPIE Proc. **2045** (1994)
Laser Processing in Manufacturing, ed. by R.C. Crafer, P.J. Oakley (Chapman
& Hall, New York 1993)
Laser Materials Processing, ed. by P. Denney, I. Miyamoto, B. Mordike (Laser
Inst. Am., Orlando, FL 1994)
S.M. Metev, V.P. Veiko: *Laser-Assisted Microtechnology*, Springer Ser. Mater.
Sci., Vol.19 (Springer, Berlin, Heidelberg 1994)
P.F. Barbara, W.H. Knox, G.A. Mourou, A.H. Zewail (eds.): *Ultrafast Phenomena IX*, Springer Ser. Chem. Phys., Vol.60 (Springer, Berlin, Heidelberg 1994)
1.2 W. Koechner: *Solid-State Laser Engineering*, 3rd edn., Springer Ser. Opt. Sci.,
Vol.1 (Springer, Berlin, Heidelberg 1992)
Solid State Lasers and New Materials. SPIE Proc. **1839** (Washington 1992)
1.3 W. Witteman: CO_2 *Lasers*, Springer Ser. Opt. Sci., Vol.53 (Springer, Berlin,
Heidelberg 1987)
1.4 Ch.K. Rhodes (ed.): *Excimer Lasers*, 2nd edn., Topics Appl. Phys., Vol.30
(Springer, Berlin, Heidelberg 1984)
1.5 M. von Allmen: In *Laser Annealing of Semiconductors*, ed. by J.M. Poate,
J.W. Mayer (Academic, New York 1982) pp.43-74
1.6 See the magazines Laser Focus (Advanced Technology, Littleton, Mass.) and
Lasers and Applications (High-Tech Publications, Torrance, Calif.)
1.7 D.C. Winburn, G. Gomez: *Practical Laser Safety* (Dekker, New York 1985)
1.8 D. Bäuerle: *Chemical Processing with Lasers*, Springer Ser. Mater. Sci., Vol.1
(Springer Berlin, Heidelberg 1986)
1.9 *Optical Thermal Response of Laser-Irradiated Tissue*, ed. by A.J. Welch, M.
van Gemert (Plenum, New York 1995)

Chapter 2

2.1 H. Kogelnik: Bell Syst. Tech. J, **44**, 455-494 (1965)
H.K.V. Lotsch: Optik **30**, 1-14, 181-201, 217-233, 563-576 (1969/70)
2.2 F. Wooten: *Optical Properties of Solids* (Academic, New York 1972)
2.3 P. Grosse: *Freie Elektronen in Festkörpern* (Springer, Berlin, Heidelberg 1979)
2.4 C.F. Bohren, D.R. Huffman: *Absorption and Scattering of Light by Small Particles* (Wiley, New York 1983)

2.5　S.A. Akhmanov, A.P. Sukhorukov, R.V. Khokhlov: Sov. Phys. – JETP **23**, 1025-1033 (1966)
S.A. Akhmanov, R.V. Khokhlov, A.P. Sukhorukov: In *Laser Handbook*, ed. by F.T. Arecchi, E.O. Schulz-Dubois (North-Holland, Amsterdam 1972) pp. 1151-1228

2.6　O. Svelto: *Progress in Optics* **12**, 3-51 (North-Holland, Amsterdam 1974)

2.7　Y.R. Shen: Rev. Mod. Phys. **48**, 1-32 (1976)

2.8　F.W. Dabby, J.R. Whinnery: Appl. Phys. Lett. **13**, 284-286 (1968)

2.9　A.I. Osipov, V.Ya. Panchenko, A.A. Filippov: Sov. J. Quantum Electron. **15**, 465-470 (1985)

2.10　E. Yablonowitch, N. Bloembergen: Phys. Lett. **14**, 907-910 (1972)

2.11　H.Y. Fan: *Semiconductors and Semimetals* **3**, Chap. 9 (Academic, New York 1967)

2.12　K.G. Svantesson, N.G. Nilsson: Phys. Scr. **18**, 405-409 (1978)

2.13　A. Bhattacharyya, B.G. Streetman: Solid State Commun. **36**, 671-675 (1980)

2.14　Y.P. Varshni: Physics **34**, 149-154 (1967)
C.D. Thurmond: J. Electrochem. Soc. **122**, 1133-1141 (1975)

2.15　I.W. Boyd, T.D. Binnie, I.B. Wilson, M.J. Colles: J. Appl. Phys. **55**, 3061-3063 (1984)

2.16　W.B. Gauster, J.C. Bushnell: J. Appl. Phys. **41**, 3850-3853 (1970)

2.17　R.G. Ulbrich: Solid-State Electron. **21**, 51-59 (1978)

2.18　E.J. Yoffa: Appl. Phys. Lett. **36**, 37-38; Phys. Rev. B **21**, 2415-2425 (1980)

2.19　A. Lietoila, J.F. Gibbons: J. Appl. Phys. **53**, 3207-3213 (1982); Appl. Phys. Lett. **40**, 624-626 (1982)

2.20　A. Haug: Solid-State Electron, **21**, 1281-1284 (1978)

2.21　A. Elci, A.L. Smirl, C.Y. Leung, M.O. Scully: Solid-State Electron. **21**, 151-158 (1978)

2.22　M. Rasolt, H. Kurz: Phys. Rev. Lett. **54**, 722-724 (1985)

2.23　V. Heine, J.A. van Vechten: Phys. Rev. B **13**, 1622-1626 (1976)
J.C. Inkson: J. Phys. C **9**, 1177-1183 (1976)

2.24　J.A. van Vechten, R. Tsu, R.W. Saris: Phys. Lett. A **74**, 422-426 (1979)
J.A. van Vechten: In *Semiconductor Processes Probed by Ultrafast Spectroscopy*, ed. by R.R. Alfano (Academic, New York 1984) Vol.2, pp.95-169

2.25　A. Compaan, A. Aydinly, M.C. Lee, H.W. Lo: *MRS Proc.* **4**, 43-48 (Elsevier, New York 1982)

2.26　G.E. Jellison, Jr., D.H. Lowndes, R.F. Wood: *MRS Proc.* **13**, 35-42 (Elsevier, New York 1983)

2.27　D.H. Auston, J.A. Golovshenko, A.L. Simons, R.E. Slusher, R.P. Smith, C.M. Murko, T.N.C. Venkatesan: Appl. Phys. Lett. **34**, 777-779 (1979)

2.28　W.P. Dumke: Phys. Lett. A **78**, 477-480 (1980)

2.29　B. Stritzker, A. Pospieszczyk, J.A. Tagle: Phys. Rev. Lett. **47**, 356-358, 1676-1677 (1981)

2.30　B.C. Larson, C.W. White, T.S. Noggle, D. Mills: Phys. Rev. Lett. **48**, 337-340 (1982)

2.31　N. Baltzer, M. von Allmen, M.W. Sigrist: Appl. Phys. Lett. **43**, 826-828 (1983)

2.32　A.M. Malvezzi, H. Kurz, N. Bloembergen: Appl. Phys. A **36**, 143-146 (1985)

2.33　M.C. Downer, C.V. Shank: Phys. Rev. Lett. **56**, 761-764 (1986)

2.34　C.V. Shank, R. Yen, C. Hirlimann: Phys. Rev. Lett. **50**, 454-457 (1983)

2.35　W.J. Siekhaus, J.H. Kinney, D. Milam, L.L. Chase: Appl. Phys. A **39**, 163-166 (1986)

2.36 P. Bräunlich, A. Schmid, P. Kelly: Appl. Phys. Lett. **26**, 150-153 (1975)
2.37 L.B. Glebov, O.M. Efimov, G.T. Petrovskii, P.N. Rogovtsev: Sov. J. Quantum Electron. **15**, 1367-1370 (1985)
2.38 N. Bloembergen: IEEE J. **QE-10**, 375-386 (1974)
2.39 H.L. Holway: J. Appl. Phys. **45**, 677-683 (1974)
      H.L. Holway, D.W. Fradin: J. Appl. Phys. **46**, 279-291 (1975)
2.40 M. Bass, H. Barrett: IEEE J. **QE-8**, 338-342 (1972)
2.41 M. Bass, D.W. Fradin: IEEE J. **QE-9**, 890-896 (1973)
2.42 N. Bloembergen: Appl. Opt. **12**, 661-664 (1973)
2.43 D. Ryter, M. von Allmen: IEEE J. **QE-17**, 2015-2017 (1981)
2.44 W.L. Smith: Opt. Eng. **17**, 489-503 (1978)
2.45 J. Bass: In *Landolt-Börnstein*, New Ser., Vol.15a, ed. by K. Hellwege, J.L. Olsen (Springer, Berlin, Heidelberg 1982) pp.5-137
2.46 A.V. Grosse: Rev. Hautes Temp. Refract. **3**, 115 – 146 (1966)
2.47 M. Sparks, E. Loh, Jr.: J. Opt. Soc. Am. **69**, 847-858 (1979)
      R.E. Lindquist, A.W. Ewald: Phys. Rev. **135**, A191-194 (1964)
      L.V. Nomerovannaya, M.M. Kirillova, M.M. Noskov: Sov. Phys. – JETP **33**, 405-409 (1971)
2.48 V.A. Batanov, F.V. Bunkin, A.M. Prokhorov, V.B. Fedorov: Sov. Phys. – JETP **36**, 311-322 (1973)
2.49 O.N. Krokhin: In *Laser Handbook*, ed. by F.T. Arecchi, E.O. Schulz-DuBois (North-Holland, Amsterdam 1972) Chap. 7
2.50 D.C. Emmony, R.P. Howson, L.J. Willis: Appl. Phys. Lett. **23**, 598-600 (1973)
2.51 J.F. Young, J.E. Sipe, J.S. Preston, H.M. van Driel: Appl. Phys. Lett. **41**, 261-264 (1982)
2.52 P.A. Temple, M.J. Soileau: IEEE J. **QE-17**, 2067-2071 (1981)
2.53 J.E. Sipe, J.F. Young, J.S. Preston, H.M. van Driel: Phys. Rev. B **27**, 1141-1154 (1983); ibid. **30**, 2001-2015 (1984)
2.54 Z. Guosheng, P.M. Fauchet, A.E. Siegman: Phys. Rev. B **26**, 5366-5381 (1982)
2.55 D.J. Ehrlich, S.R. Brueck, J.Y. Tsao: Appl. Phys. Lett. **41**, 630-632 (1982)
2.56 F. Keilmann, Y.H. Bai: Appl. Phys. A **29**, 9-18 (1982)
2.57 H.J. Leamy, G.A. Rozgonyi, T.T. Sheng, G.K. Celler: Appl. Phys. Lett. **32**, 535-538 (1978)
2.58 I.W. Boyd, S.C. Moss, T.F. Boggess, A.L. Smirl: Appl. Phys. Lett. **45** 80-82 (1984)
2.59 J.C. Koo, R.E. Slusher: Appl. Phys. Lett. **28**, 614-616 (1976)
2.60 M. von Allmen, W. Lüthy, K. Affolter: Appl. Phys. Lett. **33**, 824-825 (1978)
2.61 W.G. Hawkins, D.K. Biegelsen: Appl. Phys. Lett. **42**, 358-360 (1983)
2.62 M. Combescot, J. Bok, C. Benoit a la Guillaume: Phys. Rev. B **29**, 6393-6395 (1984)
2.63 T.R. Anthony, H.E. Cline: J. Appl. Phys. **48**, 3888-3894 (1977)
2.64 K. Affolter, W. Lüthy, M. Wittmer: Appl. Phys. Lett. **36**, 559-561 (1980)
2.65 G. Gorodetsky, J. Kanicki, T. Kazyaka, R.L. Melcher: Appl. Phys. Lett. **46**, 547-549 (1985)
2.66 N. Postacioglu, P. Kapadia, J. Dowden: J. Phys. D **24**, 1288-1292 (1991)
2.67 C. Hill: In *Laser Annealing of Semiconductors*, ed. by J. M. Poate, J.W. Mayer (Academic, New York 1982) pp.479-557
2.68 F. Keilmann: Phys. Rev. Lett. **51**, 2097-2100 (1983)
2.69 J.F. Ready: IEEE J. **QE-12**, 137-142 (1976)
2.70 D.J. Broer, L. Vriens: Phys. A **32**, 107-123 (1983)

2.71 A. Blatter, C. Ortiz: J. Appl. Phys. **73**, 8552-8560 (1993)
2.72 C. Ortiz, A. Blatter: Thin Solid Films **218**, 209-218 (1992)
2.73 D. Tuckerman, R.L. Schmitt: Proc. 1985 Multilevel Interconnect Conf. (IEEE, New York 1985) pp.24-31
2.74 A. Bächli, A. Blatter: Surface Coating Technology **45**, 393-397 (1991)
2.75 D.H. Lowndes, M. DeSilva, M.J. Godhole, A.J. Pedraza, D.B. Geohegan: Mat. Res. Soc. Proc. **285**, 191-196 (1993)
2.76 R.J. Baseman, T.-S. Kuan, M.O. Aboelfotoh, J.C. Andreshak, R.E. Turene, R.A. Previti-Kelly, J.G. Ryan: Mat. Res. Soc. Proc. **236**, 361-369 (1992)
2.77 A. Bächli, A. Blatter, M. Maillat, H.E. Hintermann: Surface Modification Technologies V, ed. by T.S. Sudarshan, J.F. Braza, The Institute of Materials 821-833 (1992)
2.78 M. Rothschild, C. Arnone, D.J. Ehrlich: J. Vac. Sci. Technol. B **4**, 310-314 (1986)
A. Blatter, U. Bögli, L.L. Bouilov, N.I. Chapliev, V.I. Konov, S.M. Pimenov, A.A. Smolin, B.V. Spytsin: Proc. Electrochem. Soc. **91**, 357-364 (1991)
2.79 B.J. Palmer, R.G. Gordon: Thin Solid Films **158**, 313-341 (1988)
2.80 V.N. Tokarev, J.L.B. Wilson, M.G. Jubber, P. John, D.K. Milne: Diamond and Related Materials (Submitted June 1994)
P. Tosin, A. Bächli, A. Blatter: Proc. Symp. on New Diamond and Related Materials, Eight CIMTEC (June 1994)
2.81 W.W. Duley, W.A. Young: J. Appl. Phys. **44**, 4236-4237 (1973)
2.82 E.V. Dan'shchikov, F.V. Lebedev, A.V. Ryazanov: Sov. J. Quantum Electron. **14**, 960-964 (1984)
2.83 R.W. Keyes: *Semiconductors and Semimetals* **4**, 327-341 (Academic, New York 1968)
2.84 G.P. Banfi, P.G. Gobbi: Plasma Phys. **21**, 845-859 (1979)

Chapter 3

3.1 R. Srinivasan, B. Braren: Chem. Rev. **89**, 1303 (1989)
3.2 R.E. Harrington: J. Appl. Phys. **38**, 3266-3270 (1967)
3.3 A.V. Benkov, A.V. Zinov'ev, T. Usmanov, S.T. Azizov: Sov. J. Quantum Electron. **15**, 643-649 (1985)
3.4 H. Salzmann: Phys. Lett. A **41**, 363-364 (1972)
R.J. Bickerton: Nuclear Fusion **13**. 457-458 (1973)
3.5 H.S. Carlslaw, J.C. Jaeger: *Conduction of Heat in Solids*, 2nd. edn. (Oxford Univ. Press, Oxford 1959)
3.6 M. Lax: Appl. Phys. Lett. **33**, 786-788 (1978)
M. Lax: J. Appl. Phys. **48**, 3919-3923 (1977)
3.7 H.E. Cline, T.R. Anthony: J. Appl. Phys. **48**, 3895-3900 (1977)
3.8 Y.I. Nissim, A. Lietoila, R.B. Gold, J.F. Gibbons: J. Appl. Phys. **51**, 274-279 (1980)
3.9 M. Sparks, E. Loh, Jr.: J. Opt. Soc. Am. **69**, 847-858 (1979)
3.10 M. von Allmen, W. Lüthy, M.R.T. Siregar, K. Affolter, M.-A. Nicolet: *AIP Proc.* **50**, 43-47 (Am. Inst. Phys., New York 1979)
3.11 E. Liarokapis, Y.S. Raptis: J. Appl. Phys. **57**, 5123-5126 (1985)
3.12 A.G. Cullis, H.C. Webber, P. Bailey: J. Phys. E **12**, 688 (1979)
3.13 Z.L. Liau, B.Y. Tsaur, J.W. Mayer: Appl. Phys. Lett. **34**, 221-223 (1979)

3.14 R.B. Gold, J.F. Gibbons: J. Appl. Phys. **51**, 1256-1258 (1980)
3.15 J.F. Gibbons, T.W. Sigmon: In *Laser Annealing of Semiconductors*, ed. by J.M. Poate, J.W. Mayer (Academic, New York 1982) pp.325-382
3.16 B.O. Boley, J.H. Weiner: *Theory of Thermal Stresses*, (Wiley, New York 1960)
3.17 G.A. Rozgonyi, H. Baumgart: J. Physique **41**, C4-5, 85-88 (1980)
3.18 Y. Matsuoka: J. Phys. D **9**, 215-224 (1976)
3.19 L. Correra, G.G. Bentini: J. Appl. Phys. **54**, 4330-4337 (1983)
3.20 H.E. Cline: J. Appl. Phys. **54**, 2683-2691 (1983)
3.21 D.M. Follstaedt, S.T. Picraux, P.S. Peercy, W.R. Wampler: Appl. Phys. Lett. **39**, 327-329 (1981)
3.22 F. Haessner, W. Seitz: J. Mat. Sci. **6**, 16-18 (1971)
3.23 L. Buene, D. Jacobson, D.C. Nakahara, J.M. Poate, C.W. Draper, J.D. Hirvonen: *MRS Proc.* **1**, 583-590 (Elsevier, New York 1981)
3.24 S.S. Lau: J. Vac. Sci. Technol. **15**, 1656-1661 (1978)
3.25 G.L. Olson, S.A. Kokorowski, J.A. Roth, L.D. Hess: *Mat. Res. Soc. Proc.* **13**, 141-154 (Elsevier, New York 1983)
3.26 A. Gat, J.F. Gibbons, T.J. Magee, J. Peng, V.R. Deline, P. Williams, C.A. Evans, Jr.: Appl. Phys. Lett. **32**, 276-278 (1978)
3.27 R.H. Uebbing, P. Wagner, H. Baumgart, H.J. Queisser: Appl. Phys. Lett. **37**, 1078-1079 (1980)
3.28 A. Lietoila, J.F. Gibbons, T.W. Sigmon: Appl. Phys. Lett. **36**, 765-768 (1980)
3.29 J.S. Williams: In *Laser Annealing of Semiconductors*, ed. by J.M. Poate, J.W. Mayer (Academic, New York 1982) pp.383-435
3.30 M.W. Geis, D.C. Flanders, H.I. Smith: Appl. Phys. Lett. **35**, 71-74 (1979)
3.31 J.F. Gibbons, K.F. Lee, T.J. Magee, J. Peng, R. Ormond: Appl. Phys. Lett. **34**, 831-833 (1979)
3.32 J.A. Roth, G.L. Olson, L.D. Hess: *Mat. Res. Soc. Symp. Proc.* **23**, 431-442 (Elsevier, New York 1984)
3.33 D.K. Biegelsen, N.M. Johnson, D.J. Bertelink, M.D. Moyer: *MRS Proc.* **1**, 487-502 (Elsevier, New York 1981)
3.34 P. Zorabedian, T.I. Kamins, C.I. Drowley: J. Appl. Phys. **57**, 5262-5267 (1985)
3.35 H.J. Zeiger, J.C.C. Fan, B.J. Palm, R.P. Gale, R.L. Chapman: In *Laser and Electron Beam Processing of Materials*, ed. by C.W. White, P.S. Peercy (Academic, New York 1980) pp.234-240
3.36 R.L. Chapman, J.C. Fan, H.J. Zeiger, R.P. Gale: Appl. Phys. Lett. **37**, 292-295 (1980)
3.37 G.H. Gilmer, H.J. Leamy: In *Laser and Electron Beam Processing of Materials*, ed. by C.W. White, P.S. Peercy (Academic, New York 1980) pp.227-233
3.38 G. Auvert, D. Bensahel, A. Perio, V.T. Nguyen, G.A. Rozgonyi: Appl. Phys. Lett. **39**, 724-726 (1981)
3.39 E. D'Anna, G. Leggieri, A. Luches: Thin Solid Films **218**, 95-108 (1992)
3.40 R.M. Walser, R.W. Bené: Appl. Phys. Lett. **28**, 624-626 (1976)
3.41 R. Pretorius, T.K. Marais, C.C. Theron: Mat. Science Eng. **10**, 1-83 (1993)
3.42 M.-A. Nicolet, S.S. Lau: *VLSI Electronics: Microstructure Sci.* **6**, Suppl.A, 329-464 (Academic, New York 1983)
3.43 T. Shibata, J.F. Gibbons, T.W. Sigmon: Appl. Phys. Lett. **36**, 566-568 (1980)
3.44 R. Andrew, L. Baufay, A. Pigeolet, L.D. Laude: J. Appl. Phys. **53**, 4862 (1982)
C. Antoniadis, M.C. Joliet: Thin Solid Films **115**, 75 (1984)
M.C. Joliet, C. Antoniadis, R. Andrew, L.D. Laude: Appl. Phys. Lett. **46**, 266 (1985)

3.45 V.A. Bobyrev, F.V. Bunkin, N.A. Kirichenko, B.S. Luk'yanchuk, A.V. Sima-
      kin: Sov. J. Quantum Electron.: **12**, 429-434 (1982)
      R. Merlin, T.A. Perry: Appl. Phys. Lett. **45**, 852-853 (1984)
      T. Szörenyi, L. Baufay, M.C. Joliet, F. Hanus, R. Andrew, I. Hevesi: Appl.
      Phys. A **39**, 251-255 (1986)
3.46 A.G. Akimov, A.P. Gagarin, V.G. Dagurov, V.W. Makin, S.D. Pudkov: Sov.
      Phys. Tech. Phys. **25**, 1439-1441 (1980);
      S.D. Pudkov: Sov. Phys. Techn. Phys. **25**, 1439-1441 (1980);
      T.E. Orlowski, H. Richter: Appl. Phys. Lett. **45**, 241-243 (1984)
      M. Thuillard, M. von Allmen: Appl. Phys. Lett. **47**, 936-938 (1985); ibid., **48**,
      1045 (1986)
3.47 I. Ursu, L. Nanu, I.N. Michailescu: Appl. Phys. Lett. **49**, 109 (1986)
      M. Thuillard, M. von Allmen: *E-MRS Proc.* **11**, 137-142 (Editions Physique,
      Les Ulis 1986)
3.48 I. Ursu, I.N. Mihailescu, A.M. Prokorov, V.I. Konov (eds.): *Laser Processing
      and Diagnostics* (Editions Physics, Les Ulis 1986) p.223
3.49 E. D'Anna, G. Leggieri, A. Luches: Thin Solid Films **218**, 219-230 (1992)
3.50 N. Bottka, P.J. Walsh, R.Z. Dalbey: J. Appl. Phys. **54**, 1104 (1983)
      G.J. Fisanik, M.E. Gross, J.B. Hopkins, M.D. Fennell, K.J. Schnoes. A. Katzir:
      J. Appl. Phys. **57**, 1139-1142 (1985)
3.51 D. Bäuerle (ed.): *Laser Processing and Diagnostics*, Springer, Ser. Chem.
      Phys., Vol. 39 (Springer, Berlin, Heidelberg 1984)
3.52 F.D. Seaman, D.S. Gnanamutu: Metal Progress (August 1975) pp.67-74
3.53 C.M. Banas:In *Physical Processes in Laser-Materials Interaction*, ed. by M.
      Bertolotti, Nato ASI Ser. B **84**, 143-162 (Plenum, New York 1983)

Chapter 4

4.1   G.A. Kachurin, N.B. Pridachin, L.S. Smirnov: Sov. Phys. Semicond. **9**, 946
      (1975)
      I.B. Khaibullin, E.I. Shtyrkov, M.M. Zaripv, R.M. Bayazitov, M.F. Galyatudi-
      nov: Rad. Eff. **36**, 225 (1978)
4.2   W.A. Elliot, F.P. Gagliono, G. Krauss: Metallurg. Trans. **4**, 2031-2037 (1972)
4.3   J. Crank: *The Mathematics of Diffusion*, 2nd edn. (Oxford Univ. Press, Oxford
      1975)
4.4   Z.L. Liau, B.Y. Tsaur, S.S. Lau, I. Golecki, J.W. Mayer: *Am. Inst. Phys. Proc.*
      50, 105-110 (AIP, New York 1979)
4.5   H.S. Carlslaw, J.C. Jaeger: *Conduction of Heat in Solids*, 2nd. edn. (Oxford
      Univ. Press, Oxford 1959)
4.6   D.H. Auston, J.A. Golovshenko, A.L. Simons, R.E. Slusher, R.P. Smith, C.M.
      Murko, T.N.C. Venkatesan: Appl. Phys. Lett. **34**, 777-779 (1979)
4.7   P. Baeri, S.U. Campisano: In *Laser Annealing of Semiconductors*, ed. by J.M.
      Poate, J.W. Mayer (Academic, New York 1982) pp.41-75
4.8   S.C. Hsu, S. Chakravorty, R. Mehrabian: Met. Trans. **9b**, 221-228 (1987)
4.9   M. von Allmen: *Mat. Res. Soc. Symp. Proc.* **13**, 691-702 (Elsevier, New York
      1983)
      A. Bächli, A. Blatter: Refr. Met. and Hard Mat. **11**, 113-119 (1992)

4.10 A.G. Cullis, H.C. Webber, J.M. Ponte, A.L. Simons: Appl. Phys. Lett. **36**, 320-323 (1980)

4.11 G.J. Galvin, M.O. Thompson, J.W. Mayer, R.B. Hammond, N. Poulter, P.S. Peercy: Phys. Rev. Lett. **48**, 33-36 (1982)

4.12 P. Schvan, R.E. Thomas: J. Appl. Phys. **57**, 4738-4741 (1985)

4.13 J.C. Baker, J.W. Cahn: In *Solidification* (ASM, Metals Park, Ohio 1971) pp.23-58

4.14 K.A. Jackson: In *Treatise in Solid State Chemistry*, ed. by N.B. Hannay (Plenum, New York 1975) Vol.5, Chap.5

4.15 F. Spaepen, D. Turnbull: *Am. Inst. Phys. Conf. Proc.* **50**, 73-83 (AIP, New York 1979)

4.16 F.F. Abraham, J.Q. Broughton: Phys. Rev. Lett. **56**, 734-737 (1986)

4.17 K.A. Jackson: Can. J. Phys. **36**, 683-891 (1958)

4.18 B. Chalmers: *Principles of Solidification* (Krieger, Huntington, Reprint 1977)

4.19 J. Frenkel: *Kinetic Theory of Liquids* (Oxford Univ. Press, Oxford 1946)

4.20 M.M. Martynyuk: Russ. J. Phys. Chem. **53**, 1080-1081 (1979)

4.21 P. Hermes, B. Danielzik, N. Fabricius, D. von der Linde, J. Kuhl, J. Heppner, B. Stritzker, A. Pospieszcyk: Appl. Phys. A **39**, 9-11 (1986)

4.22 D. Kashchiev: Surf. Sci. **14**, 209-220 (1969)

4.23 K.F. Kelton, A.L. Greer, C.V. Thompson: J. Chem. Phys. **79**, 6261-6267 (1983)

4.24 P.S. Peercy, D.M. Follstaedt, S.T. Picraux, W.R. Wampler: *Mat. Rex. Soc. Symp. Proc.* **4**, 401-406 (Elsevier, New York 1982)

4.25 W.F. Tseng, J.W. Mayer, U.S. Campisano, G. Foti, E. Rimini: Appl. Phys. Lett. **32**, 824-826 (1978)

4.26 J. Narayan, C.W. White, M.J. Aziz, B. Stritzker, A. Walthuis: J. Appl. Phys. **57**, 564-567 (1985)

4.27 G. Foti, E. Rimini, W.F. Tseng, J.W. Mayer: Appl. Phys. **15**, 365-369 (1978)

4.28 A.G. Cullis, H. Webber, N.G. Chew, J.M. Poate, P. Baeri: Phys. Rev. Lett. **49**, 219 (1982)

4.29 G.J. Galvin, J.W. Mayer: Appl. Phys. Lett. **46**, 644-646 (1985)

4.30 P.L. Liu, R. Yen, N. Bloembergen, R.T. Hodgson: Appl. Phys. Lett. **34**, 864-866 (1979)

4.31 R. Tsu, R.T. Hodgson, T.Y. Tan, J.E. Baglin: Phys. Rev. Lett. **42**, 1356-1358 (1979)

4.32 B.C. Bagley, M.S. Chen: *Am. Inst. Phys. Proc.* **50**, 97-101 (AIP, New York 1979)

4.33 E.P. Donovan, F. Spaepen, D. Turnbull: Appl. Phys. Lett. **42**, 698-700 (1983)

4.34 F. Spaepen, D. Turnbull: In *Laser Annealing of Semiconductors*, ed. by J.M. Poate, J.W. Mayer (Academic, New York 1982) pp.15-42

4.35 P.H. Bucksbaum, J. Bokor: Phys. Lett. **53**, 182-185 (1984)

4.36 P.H. Bucksbaum, J. Bokor: *Mat. Res. Soc. Symp. Proc.* **13** (Elsevier, New York 1983) pp.51-56

4.37 G. Foti, G. della Mea, E. Jannitti, G. Majni: Phys. Lett. A **68**, 368-370 (1978)

4.38 T.N.C. Venkatesan, D.H. Auston, J.A. Golovchenko, C.M. Surko: In *Laser and Electron Beam Processing of Materials* (Am. Inst. Phys., New York 1979) pp.629-633

4.39 S.U. Campisano, G. Foti, E. Rimini, F.H. Eisen, W.F. Tseng, M.-A. Nicolet, J.L. Tandon: J. Appl. Phys. **51**, 295-298 (1980)

4.40 P.A. Barnes, H.J. Leamy, J.M. Poate, S.D. Ferris, J.S. Williams, G.K. Celler: Appl. Phys. Lett. **33**, 965 (1978)

4.41 E.D. Davies, E.F. Kennedy, R.G. Ryan, J.P. Lorenzo: *Mat. Res. Soc. Proc.* **1**, 247-253 (North Holland, New York 1981)

4.42 J.S. Williams: In *Laser Annealing of Semiconductors*, ed. by J.M. Poate, J.W. Mayer (Academic, New York 1982) pp.383-435

4.43 C.W. White, S.R. Wilson, B.R. Appleton, F.W. Young, Jr.: J. Appl. Phys. **51**, 738-749 (1980)

4.44 J.M. Poate: *Mat. Res. Soc. Symp. Proc.* **4**, 121-130 (Elsevier, New York 1982)

4.45 C.W. White, J. Narayan, B.R. Appleton, S.R. Wilson: J. Appl. Phys. **50**, 2967-2069 (1979)

4.46 J.C. Baker, J.W. Cahn: Acta Met. **7**, 575-578 (1969)

4.47 S.U. Campisano, G. Foti, P. Baeri, M.G. Grimaldi, E. Rimini: Appl. Phys. Lett. **37**, 719-721 (1980)

4.48 B.C. Larson, C.W. White, B.R. Appleton, Appl. Phys. Lett. **32**, 801-803 (1978)

4.49 B.R. Appleton, B.C. Larson, C.W. White, J. Narayan, S.R. Wilson, P.P. Pronko: Am. Inst. Phys. Conf. Proc. **50**, 291-298 (AIP, New York 1979)

4.50 J.W. Cahn, S.R. Coriell, W.J. Boettinger: In *Laser and Electron Beam Processing of Materials*, ed. by C.W. White, P.S. Peercy (Academic, New York 1980) pp.89-103

4.51 K.A. Jackson, G.H. Gilmer, H.J. Leamy: In *Laser and Electron Beam Processing of Materials*, ed. by C.W. White, P.S. Peercy (Academic, New York 1980) pp.104-110

4.52 R.F. Wood: Appl. Phys. Lett. **37**, 302 (1980); and Phys. Rev. B. **25**, 2786-2811 (1982)

4.53 M.J. Aziz: J. Appl. Phys. **53**, 1158-1168 (1982); and Appl. Phys. Lett. **43** 552-554 (1983)

4.54 S.U. Campisano, D.C. Jacobson, J.M. Poate, A.G. Cullis, N.G. Chew: Appl. Phys. Lett. **46**, 846-848 (1985)

4.55 S.T. Picraux, D.M. Follstaedt: In *Surface Modification and Alloying: Aluminium*, ed. by J.M. Poate, G. Foti, D.C. Jacobson (Plenum, New York 1983) pp.288-321

4.56 J.K. Hirvonen, J.M. Poate, A. Greenwald, R. Little: Appl. Phys. Lett. **36**, 564-566 (1980)

4.57 L. Buene, J.M. Poate, D.C. Jacobson, C.W. Draper, J.K. Hirvonen: Appl. Phys. Lett. **37**, 385-387 (1980)
L. Buene, D.C. Jacobson, S. Nakahara, J.M. Poate, C.W. Draper, J.K. Hirvonen: *Mat. Ser. Soc. Proc.* **1**, 583-590 (1982)

4.58 S.T. Picraux, D.M. Follstaedt, J.A. Knapp. W.R. Wampler, E. Rimini: *Mat. Res. Soc. Symp. Proc.* **1**, 575-582 (Elsevier, New York 1981)

4.59 D.M. Follstaedt, W.R. Wampler: Appl. Phys. Lett. **38**, 140-142 (1981)

4.60 D.M. Follstaedt, S.T. Picraux, P.S. Peercy, W.R. Wampler: Appl. Phys. Lett. **39**, 327-329 (1981)

4.61 S.R. De Groot, P. Mazur: *Nonequilibrium Thermodynamics* (North Holland, Amsterdam 1962)

4.62 B.N. Bhat, R.W. Swalin: Acta Met. **20**, 1387-1396 (1972)
M. Balourdet, J. Malmejac, P. Desre: Phys. Lett. A **56**, 51-52 (1976)

4.63 L.F. Doná dalle Rose, A. Miotello: *Mat. Res. Soc. Symp. Proc.* **14** (Elsevier, New York 1982) pp.425-430

4.64 M. von Allmen, M. Wittmer: Appl. Phys. Lett. **34**, 68-70 (1979)

4.65 M. Wittmer, M. von Allmen: J. Appl. Phys. **50**, 4786-4790 (1979)

4.66 M. Wittmer, W. Lüthy, B. Studer, H. Melchior: Solid-State Electron. **24**, 141-145 (1981)

4.67 G. Badertscher, R.P. Salathé, W. Lüthy: Electron. Lett. **16**, 113 (1980)

4.68 See, e.g., E. Ben-Jacob, P. Garik: Nature **343**, 523-530 (1990)
W. Kurz, D.J. Fisher: *Fundamentals of Solidification*, 3rd edn. (Trans. Tech. Publ. 1989)

4.69 W.W. Mullins, R.F. Sekerka: J. Appl. Phys. **35**, 444-451 (1964);
R.F. Sekerka: J. Appl. Phys. **36**, 264-268 (1965)

4.70 J. Narayan: J. Appl. Phys. **52**, 1289-1293 (1981)

4.71 J. Narayan, H. Naramoto, C.W. White: J. Appl. Phys. **53**, 912-915 (1982)

4.72 M. von Allmen, S.S. Lau, T.T. Sheng, M. Wittmer: In *Laser and Electron Beam Processing of Materials*, ed. by C.W. White, P.S. Peercy (Academic, New York 1980) pp.524-529

4.73 M. Wittmer, W. Lüthy, M. von Allmen: Phys. Lett. A **74**, 127-130 (1979)

4.74 J.M. Poate, H.J. Leamy, T.T.Sheng, G.K. Celler: Appl. Phys. Lett. **33**, 918-920 (1978)

4.75 G.J. van Gurp, G.E.J. Eggermont, Y. Tamminga, W.T. Stacy, J.R.M. Gijsbers: Appl. Phys. Lett. **35**, 237 (1979)

4.76 S.S. Lau, F.W. Tseng, M.-A. Nicolet, J.W. Mayer, J.A. Minnucci, A.R. Kirkpatrick: Appl. Phys. Lett. **33**, 235 (1978)

4.77 H.J. Leamy, C.J. Doherty, K.C.R. Chiu, J.M. Poate, T.T. Sheng, G.K. Celler: In *Laser and Electron Beam Processing of Materials*, ed. by C.W. White, P.S. Peercy (Academic, New York 1980) pp.581-587

4.78 S.S. Lau, B.Y. Tsaur, M. von Allmen, J.W. Mayer, B. Stritzker, C.W. White, B. Appleton: Nucl. Inter. Meth. **182/183**, 97-105 (1981)

4.79 R.T. Tung, J.M. Gibson, D.C. Jacobson, J.M. Poate: Appl. Phys. Lett. **43**, 476 (1983)

4.80 H.E. Cline: J. Appl. Phys. **53**, 5898-5903 (1982)

4.81 M.A. Bösch, A.H. Dayem, T.R. Harrison, R.A. Lemons: Appl. Phys. Lett. **41**, 363-364 (1982)

4.82 B.M. Ditchek, T. Emma: Appl. Phys. Lett. **45**, 955 (1984)

4.83 G.G. Gladush, L.S. Krasitskaya, E.B. Levchenko, A.L. Chernyakov: Sov. J. Quantum Electron. **12**, 408-412 (1982)

4.84 T.R. Antony, H.E. Cline: J. Appl. Phys. **48**, 3888-3894 (1977)

4.85 T. Chande, J. Mazumder: Appl. Phys. Lett. **41**, 42 (1982)

4.86 J.D. Ayers, T.R. Tucker: Thin Solid Films **73**, 201-207 (1980)

4.87 D.B. Snow, E.M. Breinan, B.H. Kear: In *Superalloys* (Am. Soc. Metals, Metals Park, Ohio 1980)

4.88 C.W. Draper, L. Buene, J.M. Poate, D.C. Jacobson: Appl. Opt. **20**, 1730-1732 (1981)

4.89 H.W. Bergmann, B.L. Mordike: Z. Metallkde **71**, 658-665 (1980)
H.W. Bergmann, B.L. Mordike: Z. Werkstofftech. **14**, 228-237 (1983)

4.90 C.W. Draper, F.J.A. den Broeder, D.C. Jacobson, E.N. Kaufmann, M.L. McDonald, J.M. Van den Berg: *Mat. Res. Soc. Symp. Proc.* **4**, 419-424 (Elsevier, New York 1982)

4.91 C.W. Draper: Appl. Opt. **20**, 3093-3096 (1981)
C.W. Draper, C.A. Ewing: J. Mat. Sci. **19**, 3815-3825 (1984)
C.W. Draper, J.M. Poate: Int. Met. Rev. **30**, 85-108 (1985)

4.92 V.S. Kovalenko, V.I. Volgin: Fiz. Khim. Obrab. Mater. **3**, 28 (1978)

4.93 V.P. Greco: Plating Surf. Finishing **68**, 56 (1981)

4.94 P.G. Moore, E. Mccafferty: J. Electrochem. Soc. **128**, 1391-1393 (1981)

4.95 A.K. Jain, V.N. Kulkarni, K.B. Nambiar, D.K. Sood, S.C. Sharma, P. Mazzoldi: Rad. Eff. **63**, 175-181 (1982)
4.96 C.W. Draper, D.C. Jacobson, J.M. Gibson, J.M. Poate, J.M. Vandenberg, A.G. Cullis: *Mat. Res. Soc. Symp. Proc.* **4**, 413-418 (Elsevier, New York 1982)
4.97 C.W. Draper, L.S. Meyer, D.C. Jacobson, L. Buene, J.M. Poate: Thin Solid Films **75**, 237-240 (1981)
4.98 P. Chaudhari, B.C. Giessen, D. Turnbull: Scientific American **242**, 84-86 (April 1980)
4.99 I.W. Donald, H.A. Davies: J. Noncryst. Solids **30**, 77-85 (1978)
4.100 K. Affolter, M. von Allmen: Appl. Phys. A **33**, 93-96 (1984)
4.101 T.B. Massalsky: Proc. 4th Int. Conf. Rapidly Quenched Metals, Sendai (1981) pp.203-208
4.102 D.R. Uhlmann: J. Non-Cryst. Solids **7**, 337-348 (1972);
      H.A. Davies: Phys. Chem. Glasses **17**, 159-173 (1976)
4.103 N. Saunders, A.P. Miodownik: Ber. Bunsenges. Phys. Chem. **87**, 830-834 (1983)
4.104 D. Turnbull: Contemp. Phys. **10**, 473-483 (1969)
4.105 U. Kambli, M. von Allmen, N. Saunders, A.P. Miodownik: Appl. Phys. A **36**, 189-192 (1985)
4.106 M. von Allmen, S.S. Lau, M. Mäenpää, B.Y. Tsaur: Appl. Phys. Lett. **36**, 205-207 (1980)
4.107 M. von Allmen, S.S. Lau, M. Mäenpää, B.Y. Tsaur: Appl. Phys. Lett. **37**, 84-86 (1980)
4.108 K. Affolter, M. von Allmen, H.P. Weber, M. Wittmer: J. Non-Crystal Solids **55**, 387-393 (1983)
4.109 E. Huber, M. von Allmen: Phys. Rev. B **28**, 2979-2984 (1983); and B **31**, 3338-3342 (1985);
      U. Kambli, E. Huber, M. von Allmen: Phys. Rev. B **33**, 8643-8648 (1986)
4.110 G.G. Borodina, CH.V. Kopetskii, V.S. Kraposhin, N.V. Edneral: Sov. Phys. Dokl. **26**, 761-763 (1981)
4.111 R. Becker, G. Sepold, P.L. Ryder: Scripta Metall. **14**, 1283-1285 (1980)
4.112 H.W. Bergmann, B.L. Mordike: J. Mat. Sci. **16**, 863-869 (1981)
4.113 H. Yoshioka, K. Asami, K. Hashimoto: Scripta Metall. **18**, 1215-1218 (1984)
4.114 E.M. Breinan, B.H. Kear, C.M. Banas: Physics Today 44-50 (Nov.1976)
4.115 S. Yatsuya, T.B. Massalski: Mat. Sci. Eng. **54**, 101-111 (1982)
4.116 C.J. Lin, F. Spaepen: Appl. Phys. Lett. **41**, 721 (1982)
4.117 C.J. Lin, F. Spaepen, D. Turnbull: J. Noncryst. Sol. **61/62**, 707-712 (1984)
4.118 A. Wolthuis, B. Stritzker: J. Physique **44**, C5, 489-491 (1983)
4.119 J. Fröhlingsdorf, B. Stritzker: MRS Conf. Proc. **51**, 271-276 (Mat. Res. Soc., Pittsburg, PA 1986)
4.120 P. Mazzoldi, G. Della Mea, G. Battaglin, A. Miotello, M. Servidori, D. Bacci, E. Jannitti: Phys. Rev. Lett. **44**, 88-91 (1980);
      D.M. Follstaedt, P.S. Peercy, W.R. Wampler: Phys. Rev. Lett. **46**, 1250 (1981);
      P. Mazzoldi, G. Della Mea, G. Battaglin, A. Miotello, M. Servidori, D. Bacci, E. Jannitti: Phys. Rev. Lett. **46**, 1251 (1981)
4.121 A. Blatter, M. von Allmen: Phys. Rev. Lett. **54**, 2103-2106 (1985)

Chapter 5

5.1 R.G. Ross, D.A. Greenwood: In *Progress in Materials Science*, Vol.14., ed. by B. Chalmers, W. Hume-Rothery (Pergamon, London 1969)

5.2 V.A. Batanov, F.V. Bunkin, A.M. Prokhorov, V.B. Fedorov: Sov. Phys. JETP **36**, 311-322 (1973)

5.3 F. Hensel: In *Properties of Liquid Metals*, ed. by S. Takeuchi (Taylor and Francis, London 1973) pp.357-363

5.4 L.D. Landau, E.M. Lifshitz: *Statistical Physics* (Pergamon, London 1958)

5.5 A. Kantrowitz: J. Chem. Phys. **19**, 1097-1100 (1951)

5.6 M.M. Martynyuk: Sov. Phys. Tech. Phys. **19**, 793-797 (1974)

5.7 B. Luk'yanchuk, N. Bityurin, S. Anisimov, D. Bäuerle: Appl. Phys. A **57**, 367-374 (1993)

5.8 R. Srinivasan, B. Braren: Chem. Rev. **89**, 1303 (1989)

5.9 L.D. Landau, E.M. Lifshitz: *Fluid Mechanics* (Pergamon, London 1959)

5.10 Y.B. Zeldovich, Y.P. Raizer: *Physics of Shock Waves and High-Temperature Hydrodynamic Phenomena*, Vol.I (Academic, New York 1966)

5.11 M.C. Fowler, D.C. Smith: J. Appl. Phys. **46**, 138-150 (1975)

5.12 A.A. Boni, F.Y. Su: J. Appl. Phys. **44**, 4086-4094 (1973)

5.13 V.I. Bergel'son, A.P. Golub, T.V. Loseva, I.V. Nemchinov, T.I. Orlova, S.P. Popov, V. Svettsov: Sov. J. Quantum Electron. **4**, 704-706 (1974)
    E.V. Dan'shchikov, V.A. Dymshakov, F.V. Lebedev, A.V. Ryazanov: Sov. J. Quantum Electron. **15**, 1231-1237 (1985)

5.14 T.P. Hughes: *Plasmas and Laser Beams* (Hilger, Bristol 1975)

5.15 H.Hora, H. Wilhelm: Nucl. Fusion **10**, 111-116 (1970)

5.16 D.H. Gill, A.A. Dougal: Phys. Rev. Lett. **15**, 845-847 (1965)

5.17 C. DeMichelis: IEEE J. QE-5, 188-202 (1969)

5.18 N. Kroll, K.M. Watson: Phys. Rev. A **5**, 1883-1905 (1972)

5.19 F.W. Dabby, U.C. Paek: IEEE J. QE-8, 106-111 (1972)

5.20 S.I. Anisimov: Sov. Phys. JETP **27**, 182-183 (1968)

5.21 V.I. Igoshin, V.I. Kurochkin: Sov. J. Quantum Electron. **14**, 1049-1052 (1984)

5.22 V.A. Batanov, F.V. Bunkin, A.M. Prokhorov, V.B. Fedorov: JETP Lett. **11**, 69-72 (1972)

5.23 W.E. Maher, R.B. Hall: J. Appl. Phys. **47**, 2486-2493 (1976)

5.24 S.I. Anisimov, Y.A. Imas, G.S. Romanov, Y.U. Khodyko: *Effects of High-Power Radiation on Metals* (Nat. Technical Information Service, Springfield, VA 1970)

5.25 D.W. Gregg, S.J. Thomas: J. Appl. Phys. **37**, 2787-2789 (1966)

5.26 V.I. Mazhukin, A.A. Samokhin: Sov. J. Quantum Electron. **14**, 1608-1611 (1984)

5.27 W.W. Duley: *CO_2 Lasers, Effects and Applications* (Academic, New York 1976)
    M. Bass (ed.): *Laser Materials Processing* (North Holland, Amsterdam 1983)

5.28 D.J. Broer, L. Vriens: Phys. A **32**, 107-123 (1983)

5.29 M. von Allmen: J. Appl. Phys. **47**, 5460-5463 (1976)

5.30 M.M. Martynyuk: Sov. Phys. Tech. Phys. **21**, 430-433 (1976)

5.31 R.P. Gagliano, U.C. Paek: Appl. Opt. **13**, 274-279 (1974)

5.32 W.W. Duley, J.N. Gonsalves: Can. J. Phys. **50**, 216-221 (1972)

5.33 E. Kocher, L. Tschudi, J. Steffen, G. Herziger: IEEE J. **QE-8**, 120-125 (1972)

5.34 T. Chande, J. Mazumder: J. Appl. Phys. **56**, 1981-1986 (1984)
J. Dowden, M. Davis, P. Kapadia: J. Appl. Phys. **57**, 4474-4479 (1985)
5.35 P.G. Klemens: J. Appl. Phys. **47**, 2165-2174 (1976)
5.36 E.V. Locke, R.A. Hella: IEEE J. **10**, 179-185 (1974)
5.37 J.A. Robin, P. Nordin: J. Appl. Phys. **46**, 2538-2543 (1975)
5.38 W.W. Duley, J.N. Gonsalves: Optics and Laser Technology, 78-81 (April 1974)
5.39 R.L. Stegman, J.T. Schriempf, L.R. Hettche: J. Appl. Phys. **44**, 3675-3681 (1973)
5.40 Yu.P. Raizer: Sov. Phys. JETP **31**, 1148-1154 (1970)
5.41 A.A. Boni, F.Y. Su: Phys. Fluids **17**, 340-342 (1974)
5.42 Yu.P. Raizer: Sov. Phys. Usp. **23**, 789 (1980); and Sov. Phys. Quantum Electron. **14**, 40-45 (1984)
5.43 N.A. Generalov, V.P. Zimakov, G.I. Kozlov, V.A. Masyukov, Yu.P. Raizer: Sov. Phys. JETP **34**, 763-769 (1972)
5.44 F.V. Bunkin, V.I. Konov, A.M. Prokhorov, V.B. Federov: JETP Lett. **9**, 371-374 (1969)
E.L. Klosterman, S.R. Byron: J. Appl. Phys. **45**, 4751-4759 (1974)
5.45 J.P. Jackson, P.E. Nielsen: AIAA J. **12**, 1498-1501 (1974)
5.46 V.K. Goncharov, L.Y. Min'ko, E.S. Tyunina, A.N. Chumakov: Sov. J. Quant. Electron, **3**, 29-32 (1973)
A.P. Gagarin, V.V. Druzhinin, N.A. Raba, S.V. Maslenikov: Sov. Tech. Phys. Lett. **1**, 149-150 (1975)
5.47 J.P. Jackson, E.J. Jumper: Laser Digest (Air Force Weapons Lab., Kirtland AFB, NM 1975)
5.48 S. Marcus, J.E. Lowder, D.L. Mooney: J. Appl. Phys. **47**, 2966-2968 (1976)
5.49 J.A. McKay, J.T. Schriempf: Appl. Phys. Lett. **31**, 369-371 (1977)
J.E. Robin: J. Appl. Phys. **49**, 5306-5310 (1978)
5.50 P.E. Nielsen: J. Appl. Phys. **49**, 5306-5310 (1978)
5.51 N.N. Rykalin, A.A. Uglov, M.M. Nizametdinov: Sov. Phys. JETP **42**, 367-377 (1975)
B.M. Zhiryakov, N.I. Popov, A.A. Samokhin: Sov. Phys. JETP **48**, 247-252 (1978)
5.52 M. von Allmen, P. Blaser, K. Affolter, E. Stürmer: IEEE J. **QE-14**, 85-88 (1978)
5.53 W.E. Maher, R.B. Hall: J. Appl. Phys. **51**, 1338-1344 (1980)
5.54 Yu.P. Raizer: Sov. Phys. JETP **21**, 1009-1017 (1965)
5.55 V.I. Fisher, V.M. Kharash: Sov. Phys. JETP **55**, 439-443 (1982)
5.56 A. Pirri: Phys. Fluids **16**, 1435-1440 (1973)
5.57 P.E. Nielsen: J. Appl. Phys. **46**, 4501-4505 (1975)
5.58 B.S. Holmes, D.C. Erlich: J. Appl. Phys. **48**, 2396-2403 (1977)
5.59 J.D. O'Keefe, C.H. Skeen, C.M. York: J. Appl. Phys. **44**, 4622-4626 (1973)
C.G. Hoffman: J. Appl. Phys. **45**, 2125 (1974)
J.F. Ready: IEEE J. **QE-14**, 79-84 (1978)
5.60 E. Stürmer, M. von Allmen: ZAMP **28**, 1177-1182 (1977)
5.61 E. Stürmer, M. von Allmen: J. Appl. Phys. **49**, 5648-5654 (1978)
5.62 D.C. Hamilton, I.R. Pashby: Optics and Laser Technology 183-188 (August 1979)
5.63 R.C. Elton: *X-Ray Lasers* (Academic, Boston 1990)
5.64 O.N. Krokhin: Sov. Phys. Tech. Phys. **9**, 1024-1026 (1965)
5.65 A. Caruso, R. Gratton: Plasma Phys. **10**, 867-877 (1968)

5.66 P. Mulser: Z. Naturforsch. **25a**, 282-295 (1970)
5.67 A. Ng, D. Pasini, P. Celliers, D. Perfeniuk, L. Da Silva, J. Kwan: Appl. Phys. Lett. **45**, 1046 (1984)
5.68 V. Gupta, A.S. Argon, D.M. Parks, J.A. Cornie: J. Mech. Phys. Solids **40**, 141-180 (1992)
5.69 B. Steverding, H.P. Dudel: J. Appl. Phys. **47**, 1940-1945 (1976)
      F. Cottet, M. Hallouin, J.P. Romain, R. Fabbro, B. Faral, H. Pepin: Appl. Phys. Lett. **47**, 678 (1985)
5.70 C. Fauquignon, F. Floux: Phys. Fluids **13**, 386-391 (1970)
      J.L. Bobin: Phys. Fluids **14**, 2341-2354 (1971)
5.71 L. Spitzer: *Physics of Fully Ionized Gases*, 2nd edn. (Wiley, New York 1962)
5.72 A. Caruso: In *Laser Interactions and Related Plasma Phenomena*, ed. by H.J. Schwarz, H. Hora (Plenum, New York 1971) pp.289-305
5.73 A. Caruso, R. Gratton: Plasma Phys. **11**, 839-847 (1969)
5.74 N.G. Denisov: Sov. Phys. JETP **4**, 544-553 (1957)
5.75 V.L. Ginzburg: *The Propagation of Electromagnetic Waves in Plasmas*, 2nd edn. (Pergamon, London 1970)
5.76 J. Dawson, P. Kaw, B. Green: Phys. Fluid **12**, 875-882 (1969)
      D.V. Giovanielli, R.P. Godwin: Am. J. Phys. **43**, 808-817 (1975)
      Y.V. Afanas'ev, N.N. Demchenko, O.N. Krokhin, V.B. Rosanov: Sov. Phys. JETP **45**, 90-94 (1977)
      W.M. Mannheimer, D.G.Colombant: Phys. Fluids **21**, 1818-1827 (1978)
      R.P. Godwin: Appl. Opt. **18**, 3555-3561 (1979)
      T.P. Donaldson, J.E. Balmer, J.A. Zimmermann: J. Phys. D **13**, 12221-1233 (1980)
5.77 D. Walton: In *Nucleation*, ed. by A.C. Zettlemoyer (Dekker, New York 1969)
5.78 S. Stoyanov, D. Kashchiev: *Current Topics in Materials Science* Vol.7 (North-Holland, Amsterdam 1981) pp.69-141
5.79 S. Metev, K. Meteva: Appl. Surf. Sci. **43**, 402-408 (1989)
5.80 R. Kelly, R.W. Dreyfuss: Nucl. Instrum. Methods B **32**, 341-348 (1988)
5.81 S.I. Anisimov, D. Bäuerle, B.S. Luk'yanchuk: Phys. Rev. **43**, 12076-12081 (1993)
5.82 R.K. Singh, J. Narayan: Phys. Rev. B **41**, 8843-8859 (1990)
5.83 J.T. Cheungh, H. Sankur: CRC Crit. Rev. Solid State Mat. Sci. **15**, 63-109 (1988)
5.84 B. Luk'yanchuk, N. Bityurin, S. Anisimov, D. Bäuerle: Appl. Phys. A **57**, 367-374 (1993)
5.85 S.V. Gaponov, A.A. Gudkov, A.A. Fraerman: Sov. Phys. – Tech. Phys. **27**, 1130-1133 (1982)
5.86 D. Bhattacharya, R.K. Singh, P.H. Holloway: J. Appl. Phys. **70**, 5433-5439 (1991)
5.87 W.P. Barr: J. Phys. E **2**, 2 (1969)
5.88 H. Sankur: Thin Solid Films **218**, 161-169 (1992)
5.89 G. Hubler: MRS Bulletin (February 1992) pp.26-58
5.90 J.T. Cheung, J. Madden: J. Vac. Sci. Technol. B **5**, 705 (1987)
5.91 S. Metev: *Laser Processing and Diagnostics II*, ed. by D. Bäuerle, R.-L. Kompa, L. Laude, Proc. E-MRS **11**, 143-152 (Editions de Physique, Les Ulis 1986)
5.92 S.V. Gaponov, B.M. Luskin, N.N. Salashchenko: Sov. Phys. Semicond. **14**, 873-880 (1980)

5.93 H.S. Kwok: Thin Solid Films **218**, 277-290 (1992)
5.94 V. Craciun, D. Craciun, I.W. Boyd: Mat. Sci. Eng. B **18**, 178-180 (1993)
5.95 A. Sajjadi, I.W. Boyd: Appl. Phys. Lett. **63**, 1-3 (1993)

Appendix

A.1 *American Institute of Physics (AIP) Handbook*, 3rd. edn. (McGraw-Hill, New York 1972)
A.2 W.L. Smith: Opt. Eng. **17**, 489-503 (1978)
A.3 J. Bass: In *Landolt-Börnstein*, New Ser., Vol.15a, ed.by K. Hellwege, J.L. Olsen (Springer, Berlin, Heidelberg 1982) pp.5-137
A.4 A.V. Grosse: Rev. Hautes Temp. Refract. **3**, 115-146 (1966)
A.5 M. Sparks, E. Loh, Jr.: J. Opt. Soc. Am. **69**, 847-858 (1979)
     R.E. Lindquist, A.W. Ewald: Phys. Rev. **135**, A191-194 (1964)
     L.V. Nomerovannaya, M.M. Kirillova, M.M. Noskov: Sov.Phys. – JETP **33**, 405-409 (1971)
A.6 *Handbook of Chemistry and Physics*, ed. by R. Weast (CRC, Cleveland 1981) pp.55-60
A.7 Metals Reference Book, 4th edn., ed. by C.J. Smithells (Butterworth, London 1967) Vols.II-IV
A.8 A.N. Nesmeyanov: *Vapor Pressure of the Chemical Elements* (Elsevier, Amsterdam 1963)
A.9 H.S. Carlslaw, J.C. Jaeger: *Conduction of Heat in Solids*, 2nd. edn. (Oxford Univ. Press, Oxford 1959)

Subject Index

Printing: Mercedesdruck, Berlin
Binding: Buchbinderei Lüderitz & Bauer, Berlin

Springer Series in *Materials Science*

Advisors: M. S. Dresselhaus · H. Kamimura · K. A. Müller
Editors: U. Gonser · R. M. Osgood, Jr. · M. B. Panish · H. Sakaki
Managing Editor: H. K. V. Lotsch

Springer-Verlag
and the Environment

We at Springer-Verlag firmly believe that an international science publisher has a special obligation to the environment, and our corporate policies consistently reflect this conviction.

We also expect our business partners – paper mills, printers, packaging manufacturers, etc. – to commit themselves to using environmentally friendly materials and production processes.

The paper in this book is made from low- or no-chlorine pulp and is acid free, in conformance with international standards for paper permanency.